A Magnetic Spark: The Therapeutic Potential of Transcranial Magnetic Stimulation

Parekh

Contents

1 Introduction

1.1 What is Transcranial Magnetic Stimulation?

Prior to the 1980s, brain stimulation methods involving transcranial electrical stimulation were a rather painful experience [1]. Transcranial magnetic stimulation (TMS) introduced a relatively pain-free modality, enabling application of higher intensities of electrical energy across the scalp, traversing the skull and into cortical brain regions [2, 3]. Devoid of any electrically conducting units in contact with the scalp, TMS involves passing a large electric current through a coil of wire for a brief duration. As per Faraday's law, this briefly changing electric current produces a magnetic flux perpendicular to the plane of the coil. The magnetic flux penetrates across the biological tissue and induces electric current in the underlying population of neurons [4].

In their pioneering study, Barker et al. [2] used a circular coil for stimulation of the motor cortex in which the intensity of the magnetic flux and the induced current is highest at the centre of the coil. Since then, further TMS coil designs have allowed for either focal stimulation of cortical targets or stimulation of deeper regions of the brain. One of the most commonly used TMS coils, referred to as the "figure-of-eight" coil, facilitates highly focal stimulation of only superficially accessible cortical brain regions [5, 6]. It consists of two circular wire coils placed adjacent to each other; current passing through it generates a strong focal magnetic flux at the mid-point between the centres of the two coils [7]. An alternative design is the double cone coil, where the two wire coils are tilted towards each other at an obtuse angle, reaching greater depth but at the expense of reduced focality [7, 8]. A modelling study of these two coils showed that the double cone coil magnetic flux reaches a depth of ~2 cm spanning approximately 30 cm^2, whereas the figure-of-eight coil reaches a depth of ~1.4 cm, spanning only ~11 cm^2 [5]. Another novel coil for rTMS is the H-coil, which allows stimulation of deeper brain regions without inducing significant activation of cortical regions, often referred to as Deep TMS [9]. The deeper penetration of the electric flux is achieved by less focal stimulation [9].

The ease and painless application of single pulses combined with simple and straightforward interpretation of results make single pulse TMS an attractive tool for neurologists, playing a valuable role in the diagnosis of certain neurological disorders. By applying single pulses at the motor cortex to activate target muscles, clinicians can determine the conduction time for the signal to travel from the motor cortex to these muscles, referred to as the central motor conduction time (CMCT). CMCT is known to be altered in diseases involving structural lesions in conduction pathways or demyelination [10]. Measuring the alterations in CMCT using single pulse TMS has clinical prognostic value for diseases such as multiple sclerosis, amyotrophic lateral sclerosis, atrophy or palsy [11]. With the increasing promise of TMS for research and clinical/prognostic purposes, the stimulation devices became more sophisticated to deliver TMS pulses in a repetitive manner. This ushered in the era of repetitive TMS (rTMS). The merits of rTMS lay in its effects lasting beyond the duration of stimulation [7]. The mechanisms behind longer lasting effects of rTMS is likely analogous to mechanisms underlying long-term synaptic plasticity, broadly categorized into long-term potentiation (LTP) and long-term depression (LTD) [12]. LTP is the phenomenon where repeated high frequency stimulation of a pre-synaptic neuron leads to an enhanced excitatory post-synaptic potential (EPSPs) whereas LTD is the opposite phenomena of LTP. Whilst LTP strengthens synaptic connections, LTD acts to weaken synaptic connections. The understanding of LTP and LTD processes in mammalians is largely derived from studies of hippocampal slices from animals. Hill (1978) showed that rat hippocampal neurons, when put in a new environment, fire at about 5-7 Hz [13]. When these hippocampal neuron slices are electrically stimulated with 5, 10 or 20 bursts at an interval of 200 milliseconds [each burst consisting 4 pulses delivered at 100 Hz], LTP is consistently induced in the stimulated population of neurons [14, 15]. Notably, a 10 bursts stimulation caused LTP in 100% of neurons [14]. LTP results in strengthened synaptic connections between the pre-synaptic neuron and post-synaptic neuron. This implies that a single stimulus to the pre-synaptic neuron would now elicit a much higher response from the post-synaptic neuron than before. Furthermore, while LTP is caused by high frequency stimulus to the pre-synaptic neuron over a short time period, LTD is a result of slow, low frequency stimulation over a longer period. To mimic such LTP like effects over the human cortex, researchers translated the parameters from the stimulation of hippocampal neuron slices for human rTMS. A study with electroencephalogram (EEG) in humans showed that subjects engaged in a word memory task exhibited increased theta power (4-7 Hz) during encoding of words that were later successfully recalled [16]. Keeping in mind the technological limitations of the TMS stimulators at the time e.g. overheating, the parameters were re-adjusted to deliver 3 (instead of 4) pulses at 50 Hz (instead of 100 Hz), with each burst repeated every 200 milliseconds [17]. This gave rise to a form of rTMS called theta burst stimulation (TBS).

Measurements of motor evoked potential (MEP) have led to an understanding of the excitatory/inhibitory nature of rTMS. The MEP acts as an immediate readout of underlying cortical excitability. It results from action potentials (AP) generated in the interneurons which stimulate pyramidal neurons, further activating the corticospinal tract. This activates the spinal motor neurons that relay the signal to the muscle [18]. Electromyography (EMG) measures and quantifies the amplitude of a muscle twitch as a MEP [19]. This form of signal transduction resulting in muscle activation is usually referred to as the indirect volley (I-waves), as the pyramidal neurons are not directly stimulated [20]. However, higher intensity TMS pulses can result in direct simulation of the pyramidal neurons which results in direct volley of signal transduction (D-waves) [20]. The direction of the induced current and the intensity of the TMS pulses influence the amplitude of the MEP. MEPs can be utilized as a measure of motor output control, the excitability of the motor cortex or changes outside of the motor cortex that cause alterations within the motor cortex [21]. Huang et al. [22] delivered TBS continuously for 40 seconds (continuous TBS – cTBS) over the motor cortex and reported a reduction in MEP. This suggests a decrease in cortical excitability – an LTD-like phenomenon. Inversely, delivering TBS in an intermittent fashion, i.e. when 10 bursts of TBS are succeeded by a 8-second period of rest (intermittent TBS – iTBS), led to an increase in MEP, suggesting an increase in cortical excitability (LTP-like phenomena)[22]. These two forms of TBS are commonly used today to induce inhibitory (cTBS) and excitatory (iTBS) aftereffects over cortical areas [22, 23]. An important point to consider is that typical cTBS and iTBS protocols deliver the same amounts of pulses (600), however cTBS delivers this energy over a shorter span of time (40 seconds), while iTBS delivers this over a longer duration (200 seconds). Thus, the advent of TBS has allowed longer lasting effects beyond the stimulation with reduced stimulation duration and fewer number of pulses.

Studies stimulating the motor cortex and measuring the MEPs, have established that lower frequency rTMS ($\leq$ 1 Hz) results in a decrease in MEP amplitude [24] while high frequency ($\geq$ 5 Hz) causes an increase in MEP amplitude [7, 25]. This simple rule of thumb helps inform rTMS parameter selection for both research and clinical application. For example, when attempting to inhibit the activity of a cortical area, low frequency rTMS would be used but when facilitation of the underlying cortical area is desired, high frequency stimulation will be employed. However, there are exceptions to this rule. RTMS exerts its effects via excitation and inhibition of populations of neurons. It affects cellular signalling cascades and neurotransmission. The large-scale effects of rTMS is thus, a culmination of complex patterns of excitation and inhibition of individual neurons. Depending on the cortical area stimulated and the underlying brain state, the observed effects of rTMS might differ from that expected [6], as small changes in rTMS parameters can result in variable cumulative changes at the brain organ level, resulting in observations contrary to expectations. Be that as it may, the advancement of rTMS has offered researchers an opportunity to probe brain functions by non-invasively activating/inhibiting brain regions.

1.2   Cellular mechanisms of (r)TMS

Before detailing the mechanism of rTMS, we begin with the effect of single pulse TMS at the cellular level AP. Neurons at rest (i.e. no activity) maintain an electric potential difference across the neuronal membrane, with the help of trans-membrane protein transporters. These transporters exchange sodium ions for potassium ions at the expense of energy. This maintains a higher potassium concentration inside of the neuron and a higher sodium concentration outside of the neuron. Thus, the neuronal membrane maintains an electric potential (called as the resting membrane potential) of approximately -70 mV at equilibrium [26]. A depolarizing current stimulus above a certain threshold results in the opening of voltage gated sodium channels (VGNaCs). This opening of VGNaCs produces further influx of sodium ions, which in turn prompts further depolarization resulting in more VGNaCs to open. This causes a rapid depolarization until the membrane potential is equal to the equilibrium potential of sodium ions. The depolarizing current stimulus also causes slower acting voltage gated potassium ion channels (VGKCs) to open up, resulting in an efflux of potassium ions [26].

The VGNaCs close quicker than VGKCs, due to the slow dynamics of VGKCs. This delay results in a significant efflux of potassium ions from the neuron, causing repolarization of the neuronal membrane beyond the resting membrane potential. As the VGKCs close the membrane potential returns to its baseline at the expense of the Na+/K+-ATPase pump. This entire cycle of depolarization and repolarization constitutes an AP. The AP travels along each neuron via the voltage gated channels across the axon and terminates at the synapse. A simple synapse consists of two neurons, the pre-synaptic and the post-synaptic neuron. The APs in the pre-synaptic neuron result in a chemical signal cascade across the synapse to generate APs in the post-synaptic neuron. As the AP reaches the pre-synaptic terminal, it causes voltage gated Calcium channels (VGCaCs) at the pre-synaptic terminal to open. Calcium influx results in the release of neurotransmitters in the space between two neurons called a synaptic cleft. The neurotransmitters bind to ionotropic or metabotropic receptors at the post-synaptic neuron. The binding of neurotransmitters to ionotropic receptors results in opening of ion channels, giving way to either an inhibitory (further polarization) or excitatory (depolarization) response in post-synaptic neuron [26]. Metabotropic receptors, on the other hand, do not directly result in opening or closing of ion channels. Instead, the binding of neurotransmitters to metabotropic receptors leads to activation of other downstream proteins (e.g. G-proteins) that result in opening of ion channels away from the synaptic membrane. The description of AP presented is a simplistic one as the mammalian neuron AP is known to be more dynamic and complex. It is generated by a larger number of voltage-gated channels responsible for different ions. However, the outcome of a single TMS pulse delivered over a population of neurons can be understood as an electric current delivered to neurons that results in APs [26, 27].

The immediate outcome of single pulse TMS is often quantified using MEPs and motor threshold (MT). MT is commonly defined as the minimum stimulator strength required to elicit a MEP response of greater than 50 mV in the target muscles at least five out of ten times. The MT measured using single pulse TMS is dependent on sodium channel activation in cortical-cortical neurons and their synaptic connections to cortico-spinal neurons [28]. Thus, substances blocking sodium channel activation or non NMDA glutamate receptors (responsible for quick excitatory synaptic transmission) influence the MT measures [29]. MEP, like MT, is also dependent upon sodium channel activation [30]. It is hypothesized that a reduction in MEPs, suggesting reduced cortical excitability, stems from sodium channel inactivation resulting in reduced I-waves [31]. In addition, substances that modulate the inhibitory and excitatory properties of neural circuits also influence the MEP measurement [28]. It is also important to note that resulting electric current induced by single pulse TMS depends on other factors like the excitability of the neuronal elements, the orientation of excitable neurons with respect to the TMS coil, skull thickness, skin impedance, distance between skull and cortex (cerebrospinal fluid), and the parameters of TMS itself such as the intensity, frequency, duration, pulse shape and form [29] etc. This was demonstrated in a study evaluating the MEPs influences of iTBS stimulation of the motor cortex [32]. The study reported dose dependence of the response to iTBS, whereby larger doses of iTBS (more pulses, longer duration) resulted in greater excitatory effects on the motor cortex. However, this was only true for stimulation up to certain intensities. For example, when stimulated at intensities greater than 120% of resting MT (RMT: MT when the hand muscles are relaxed), all doses of iTBS (600 versus 1200 versus 1800) failed to facilitate MEPs. However, when stimulated at 90% of the RMT, the dose dependency was most clearly visible. Similarly, another study showed maximal inhibitory effects from cTBS when delivered at greater than 150% RMT. While, excitatory effects from iTBS were greatest when stimulated at 110% RMT [33].

Besides single pulse TMS, longer-term effects of rTMS on cortical excitability are mediated by mechanisms similar to LTP and LTD [34]. LTP is induced by high frequency stimulation of the pre-synaptic neuron and is mediated by the NMDA and AMPA receptors. During normal resting potential, the NMDA receptors are blocked by magnesium ions. The NMDA receptors open when a) glutamate is bound to the receptor and b) the post-synaptic neuron is depolarized. At low frequency stimulation of pre-synaptic neuron, the stimulus is not large enough to unblock NMDA receptors. EPSP is mediated only via AMPA receptors [35]. When the pre-synaptic neuron receives a high frequency stimulus, it results in a larger EPSP. The larger EPSP depolarizes the post-synaptic neuron thus unblocking the NMDA receptors. This leads to calcium ion influx in the post-synaptic neuron. The influx of calcium ions results in recruitment of additional AMPA receptors to the post-synaptic terminal. The calcium ions also activate post-synaptic signalling pathways that result in further recruitment of AMPA receptors to the post-synapse. This results in additional sensitivity to glutamate post LTP, leading to stronger post-synaptic response towards a normal pre-synaptic input [35]. These processes happen over a time frame of 1 to 2 hours. However, a slower process, called as the late phase, occurs over a longer duration due to changes in gene expression and protein synthesis. The late phase is mediated via protein kinases that cause phosphorylation of other proteins and ultimately influence transcriptional factors and gene expression possibly leading to newer synapse formation [35]. LTD on the other hand is opposite of LTP. Low frequency stimulus to pre-synaptic neuron usually leads to LTD. LTP and LTD processes nullify each other's effects when they succeed one another. The main difference then appears to be the rate at which calcium ions concentration changes in response to high frequency versus low frequency stimulus. While high frequency stimulus results in a rapid rise in calcium ion concentration on the post-synaptic terminal, low frequency stimulus leads to a slow and rather small rise in calcium ion concentration. This slow rise of calcium ions leads to activation of phosphatases – proteins that dephosphorylate other proteins. The dephosphorylation prompts a signalling cascade that ultimately results in internalization of AMPA receptors from the post-synaptic terminal. Thus, LTD results in a decrease of AMPA receptors at the post-synaptic membrane. This highlights the opposite nature of the two phenomena, and the selective role they play in retaining new information or strengthening/discarding old ones [35].

Another process that underlies the brain's response to newer information is called metaplasticity. Metaplasticity refers to the property by which the history of synaptic activity impacts the synapse's response to subsequent stimulation [34, 36, 37]. Such processes indicate that the response of neuronal populations to rTMS is likely more complex than LTP and LTD mechanisms gleaned from studies of hippocampal neurons. Many studies have employed magnetic stimulation in animal models as well as in humans as an attempt to further understand the phenomenon of synaptic plasticity underlying rTMS. Ma et al. [38] showed that low frequency magnetic stimulation of hippocampal neuron cultures led to changes in synapse density and synapse structure integrity. They showed that low intensity 1 Hz stimulation can lead to increase in synapse density and axonal arborization, while high intensity 1 Hz stimulation can lead to a detrimental effect on synapse morphology. Effects on the post-synaptic terminal in the form of increased AMPA receptors were seen in response to 10 Hz repetitive magnetic stimulation in mouse hippocampal neurons [39]. Differential effects of high and low frequency rTMS have also been reported through animal studies. Gersner et al. [40] applied multiple sessions of 20 Hz rTMS to one group of rats and 1 Hz rTMS to a second group of rats. Their results indicate that high frequency rTMS, but not low frequency rTMS may lead to changes in glutamate receptor subunit of AMPA receptors. Additionally, they compared the effects of high frequency rTMS (HF-rTMS) in awake and anesthetized animals and found that 20 Hz rTMS only increased the markers of neuroplasticity in awake animals, hinting towards the importance of underlying neuronal activity during rTMS. Another study showed that iTBS resulted in both immediate and prolonged reduction in numbers of cortical cells expressing parvalbumin but cTBS only caused such changes transiently [41]. Their results suggest different molecular mechanisms of long- and short-term plasticity for different rTMS protocols [34]. Another study [42] showed that 5Hz rTMS of human cells led to phosphorylation of CREB proteins (cAMP response element binding proteins that are transcriptional factors initiated by cAMP). CREB further modifies gene expression allowing more AMPA receptors at the post-synaptic neurons. Their study postulates that rTMS likely interacts and influences gene expression via CREB. Some well described and studied genes whose expressions are modified in response to changing calcium ion concentration have also been shown to have increased expression in response to 1 Hz, 10 Hz and iTBS protocols in rats [34]. For example, expression of genes like c-Fos and zinc finger protein 268 in certain regions of the animal brain is differentially influenced by rTMS protocols [43]. These results highlight the complexity with which the rTMS protocols may interact with large scale neuronal organization.

In addition to molecular and gene expression effects, rTMS also causes changes in neurotransmission per se [34]. Of relevance are studies with human subjects combining rTMS and single photon emission computed tomography (SPECT). One such study investigating patients with clinical depression found that 10 Hz rTMS modulated serotonin receptor binding sites – resulting in receptor increase in the dorsolateral prefrontal cortex (DLPFC) but receptor decrease in the hippocampus [44]. Further studies with human subjects show increases in dopamine in striatum in healthy subjects as well as patients with depression [45, 46]. The GABAergic and glutamatergic transmissions are also influenced by magnetic stimulation in rat brains [47]. This in turn can lead to changes in protein expression of cortical inhibitory interneurons, which then modulate the plasticity of a cortical area. While rTMS affects diverse molecular processes, Romero et al. [48] have shown that the area covered by the rTMS magnetic field, even when as small as 2 mm in diameter affects the activity of a large number of neighbouring cells. They reported varied combinations of excitation and inhibition of the neighbouring cells *in vivo* in rhesus monkeys. Thus, the overall effect of rTMS would be a culmination of such single cell effects. The choice of rTMS parameters would allow either excitatory or inhibitory processes to win over; resulting in a macro-effect that is usually reported in human rTMS studies. To produce the desired macro-effects, the parameter space for designing adequate rTMS protocols is large and diverse. The specifications that can be adjusted include total number of pulses delivered per session (e.g. 600 and 1800 for iTBS or 3000 for 10 Hz), frequency of pulse delivery (e.g. 10 Hz, 20 Hz etc.), length of the pulse train (e.g. 2 seconds, 4 seconds etc.), interval between the trains of pulses (Inter-train interval, e.g. 8 seconds, 26 seconds, 25 seconds etc.), number of sessions per day (e.g. 1-5 sessions/day), total number of sessions (20-30 total sessions), and the titrated strength of the RMT at which the stimulation is delivered (e.g. 80%, 110% or 120% of the RMT). Such a wide array of choices has led to research studies using different combinations of parameters, thus making universal claims about rTMS effects difficult. However, available published guidelines [8, 49, 50] and acceptance and initiative by the FDA have to large extent helped standardize rTMS parameters for clinical use. Regardless, rTMS with varied parameters have been crucial to study basic cognitive faculties of the human brain, leading to insights regarding the roles of different brain regions in cognitive processes.

1.3   TMS in cognitive and behaviour studies

The study of the underlying brain function with application of rTMS involves two significant methods. The first method concerns disrupting the functioning of certain brain regions by using particular parameters of rTMS. It assesses the causal effects of rTMS on cognitive abilities, thus highlighting the brain region's role in cognitive processes [51]. Such an approach is referred to as the virtual lesion approach [52]. Using such approaches, it is also possible to study the temporal dynamics of cognition i.e. the timeline by which cognitive stimuli are processed [53–55]. For example, 5 Hz rTMS at the DLPFC results in more frequent errors by subjects in a delayed match to sample task [56] and 5 Hz rTMS at parietal cortex resulted in reduced reaction times even though it did not influence the error frequency [57]. Some studies have also reported enhancement of task performance from rTMS applied before the execution of the task [58–61]. For example, Tegenthoff et al. [59] showed increased sensitivity of tactile cues up to 2 hours after 5 Hz rTMS to right index finger representation on the cortex, and Boyd and Lindell [60] showed 5 Hz rTMS to premotor cortex facilitated motor memory consolidation beyond the end of the stimulation. Several other studies have investigated brain regions using single session rTMS to understand visual attention, mental rotation, working memory, motion detection, discrimination etc. [61]. Such single session rTMS can have acute aftereffects on the brain functioning, likely through mechanisms analogous to LTP or LTD [62]. In addition, cognitive processing has also been shown to be affected by single pulse TMS. Abrahamyan et al. [63] showed a dependency of visual threshold on the intensity of pulses, wherein low intensity pulses cause an increase in visual thresholds, while high intensity pulses cause a decrease. Their results underscore the complexity of TMS techniques, showing how not only the frequency of stimulation (low versus high frequency) or the pattern of stimulation (iTBS versus cTBS), but even the intensity of stimulation can result in contradictory effects. Aside from the lesion approach, the second method is a novel technique to study brain functioning involving entrainment of underlying neuronal oscillations using rTMS [64]. Studies involving EEG in conjunction with rTMS have shown that the aftereffects of the stimulation are not only dependent upon the parameters of the rTMS (frequency, duration etc.) but also on the activation level of the brain region receiving the stimulation [65]. For example, Romei et al. [66, 67] have shown that the response of TMS in eliciting phosphenes was dependent upon the alpha band oscillatory activity of the visual cortex. Bestmann et al. [68] also reported that the activity of contralateral brain regions is influenced by the stimulation of ipsilateral brain regions in a state dependent manner. These reports on the relationship between the pre-stimulation brain state and the outcome of stimulation have led to the development of brain state dependent brain stimulation (BSDBS). Bergmann presents a conceptual framework of open and closed looped BSDBS [69]. An open loop BSDBS system identifies the underlying brain state and accordingly adjusts the timing of the rTMS delivery to stimulate during the most optimal time frame (high alpha oscillation or synchronized with alpha waves etc.). Closed loop BSDBS, on the other hand works on similar principle as the open loop BSDBS, but it additionally aims to modulate the underlying brain activity thus allowing it to either maintain or disrupt the function of brain regions. Thus, entrainment exploits the synchronous firing of neuronal populations [70]. By phase locking the rTMS frequency with the intrinsic activity of the brain region under investigation, it is possible to modulate the brain region's functionality by either enhancing the neuronal oscillations or counteracting the ongoing oscillations [64]. This novel technique holds promise for application in psychiatric disorders e.g. depression. Depression is often characterized by significant cognitive deficits across memory, attention, and learning [71–73]. A recent meta-analysis of the cognitive effects of rTMS in patients with depression found only modest enhancements resulting from multiple sessions of rTMS [62]. Thus, use of personalized rTMS approaches depending on underlying brain states such as entrainment may improve the enhancing effect of rTMS on cognitive faculties. The important takeaway though is that an extensive body of research substantiates the ability of rTMS to influence brain circuits, resulting in either improvement or deterioration of cognitive performance and behavioural output. The application of rTMS to regulate increased or decreased activity/connectivity of certain brain regions in psychiatric disorders is an extension of this premise.

1.4 RTMS as treatment modality of major depressive disorder

1.4.1 The need for rTMS in the treatment of major depressive disorder

The World Health Organization (WHO) ranks depression as one of the leading causes of disability, with its lifetime prevalence ranging between 10% - 20% of the general population [74, 75]. Depression is frequently associated with poor health, smoking habits and cardiac problems [76]. Aside from the economic costs, it also weighs adversely on interpersonal relationships. Depression is a recurring disorder characterized by emotional dysregulation [77]. It often results in a negative bias emerging as an inability to disengage from processing of negative information from the surrounding as well as from the memory [78], frequently precipitating a negativistic perception of the world, one's own self and the future. Depression is clinically characterized by an accumulation of symptoms across multiple domains: emotional, psychomotor, neurovegetative, and cognitive. These symptoms can include depressed mood, anhedonia, psychomotor agitation, fatigue, weight/appetite loss or gain, insomnia/hypersomnia, suicide ideation, poor concentration, and feelings of worthlessness and guilt [78, 79]. The Diagnostic and Statistical Manual of Mental Disorders - 5 (DSM5)[80] requires five or more of these symptoms for a diagnosis, with depressed mood or loss of interest/pleasure as one of the primary symptoms. Further, these symptoms must have persisted for two weeks or longer and significantly affected day to day life in order to make a diagnosis of depression [74]. The International Classification of Disease, 10th edition (ICD-10)[81] has a comparable diagnosis for depression. Differently from DSM5, ICD-10 requires only four or more symptoms to be present for more than 2 weeks with significant effect on daily life, with at least two core symptoms amongst depressed mood, anhedonia, and fatigue.

The first line of treatment against depression includes cognitive behaviour therapy or/and pharmacological intervention [82]. One of the main difficulties in treating depression is the interindividual variability and its varying symptom profile. Two separate patients given the diagnosis of depression might not have comparable symptoms [83]. The large number of combinations of symptoms that can be diagnosed as a depressive episode are indicative of the significant underlying variability of depression pathophysiology [84]. Considering the complexity of the disease, approximately a third of patients with depression do not respond to multiple rounds of pharmacotherapy and exhibit treatment resistant depression (TRD)[85]. Due to its accessibility for rTMS and its connection to deeper brain structures involved in emotional processing, DLPFC rTMS proved itself as a promising alternative for TRD treatment owing to its ease of application (no secondary medication needed e.g. muscle relaxants or sedation in case of electro-convulsive therapy (ECT)) and non-invasive nature (no surgical procedure required like in deep brain stimulation (DBS))[86–88]. In addition, applying rTMS at the DLPFC was also instigated because depression was associated with significant left-right prefrontal asymmetry in glucose metabolism which resolved with successful treatment [89]. 18F-fludeoxyglucose positron emission tomography (FDG-PET) studies showed depression to be associated with hypometabolism of left prefrontal regions [90] and hypermetabolism of the contralateral regions [91]. Hence, it was hypothesized that this hypoactivity and hyperactivity in the prefrontal regions could be "corrected" using HF-rTMS over the left prefrontal region and low frequency rTMS (LF-rTMS) over the right side, respectively. Thus, apart from cortical accessibility and connections to deeper regions involved in emotional processing, the hypo-activity in frontal regions observed in depression patients also influenced the choice of DLPFC as a target for rTMS stimulation [89, 90].

1.4.2 Evidence for antidepressant effects of rTMS

A string of randomized control trials (RCT) over the last three decades have confirmed the efficacy of DLPFC rTMS for the treatment of depression. O'Reardon et al. [86] and George et al. [87] have shown that 10 Hz rTMS over the left DLPFC is an efficient treatment modality for depression. Studies demonstrating the benefits of HF-rTMS eventually led to Foods and Drug Administration (FDA) approval of the clinical use of 10 Hz rTMS for depression treatment. Recent consensus studies [8, 49] have revealed that HF-rTMS over the left DLPFC has a level A recommendation implying a definite effect of left DLPFC HF-rTMS for treating depressive symptoms. LF-rTMS over the right DLPFC is also effective in the treatment of depression [92, 93], albeit with smaller effect sizes compared to HF-rTMS. The guidelines from the TMS community recommend right DLPFC LF-rTMS as a level B treatment implying there is a probable antidepressant effect of such treatment regimen but more randomized controlled studies with larger sample sizes are needed for further establishing the claim [8, 49].

Even though the clinical efficacy in the treatment of depression is consistently reported in the literature, the 10 Hz protocol has helped only 30-50% of the cases [94]. Moreover, a single session of 10 Hz rTMS delivering a total of 3000 pulses involves 75 trains (each train is 4 seconds long followed by 26 seconds break) spanning 37.5 minutes. Consequently, this prompted the move toward optimization of rTMS protocols to enhance the response rates in patients with depression. ITBS protocol came as an alternative to HF rTMS. One crucial benefit of TBS is the short duration within which excitatory effects similar to traditional rTMS protocols can be induced in the brain [8]. In fact, TBS has shown similar antidepressant effects after stimulation with greatly reduced session duration and number of pulses delivered [88]. From an economic perspective, shorter duration rTMS protocols with similar efficacy would be more advantageous for clinics, both in terms of reduced working hours for treatment delivery, more patients treated in a day and reduced wear and tear of stimulation devices [95].

Compared to 37.5 minutes of 10 Hz rTMS, several studies have reported significant antidepressant effects from much shorter TBS protocols [96–99]. However, conclusive evidence for the efficacy of iTBS came from the work of Blumberger et al. [88]. With a large sample size, the study showed that patients receiving daily iTBS of 3 minutes over the left DLPFC for four weeks, showed similar reduction in depressive symptoms as those receiving 37.5-minute 10 Hz rTMS. Since then the FDA has also approved iTBS for the clinical treatment of depression. Aside from the DLPFC as a target, a chart review by Bakker et al. [100] found no differences in safety, tolerability, or effectiveness of 6-minute iTBS versus 30-minute 10 Hz rTMS over the dorsomedial prefrontal cortex (DMPFC). While more randomized controlled studies are needed to establish the efficacy of iTBS over the DMPFC, it nevertheless establishes similar antidepressant effects from stimulation at a target other than the DLPFC. Apart from iTBS, 18 Hz Deep rTMS at the left DLPFC with the H-coil has also been tested as alternative treatment in two studies [101, 102]. Both studies reported positive effects from stimulation spanning twenty sessions. However after 20 sessions of Deep rTMS, one study reported a higher remission rate only in real condition but comparable symptom reduction across both real and sham conditions [102]. Although Deep rTMS is still not an FDA approved treatment, but the response and remission rates reported in both studies mentioned above are comparable to those from 10 Hz rTMS and iTBS.

While iTBS delivery is clearly shorter in duration with similar antidepressant effects, daily application over a span of four weeks still requires patients to visit the stimulation clinic for the entire duration. To further expedite the antidepressant effect and reduce the time encumberment of patients as well as the clinical staff, accelerated protocols have been explored [103]. A meta-analysis of both open label and randomized control trials, revealed good effect sizes from accelerated rTMS protocols [104]. Although different studies have applied a varied number of stimulation sessions per day. Some studies have delivered up to 7 sessions of rTMS a day [105], others have delivered twice daily sessions [106–109], thrice daily sessions [110], while most have delivered 5 sessions per day [111–113] with the inter-session interval varying between 15-60 minutes. The importance of applying accelerated rTMS using multiple sessions rather than a single but much longer rTMS session was highlighted by the following two studies. The first study reported on reverse effects of prolonged iTBS and cTBS, which were twice as long as the traditional 3-minute iTBS and 40-second cTBS protocols [114]. They reported that when iTBS/cTBS was delivered in twice the normal dose, the effects on MEPs were opposite i.e. longer iTBS caused inhibitory effects while the longer cTBS resulted in excitatory effects on the MEP. Furthermore, a retrospective chart study analysed the differences in rTMS antidepressant action in two groups of patients [115]. One group received a single session of 10 Hz rTMS in both the hemispheres (3000 pulses in each of the two hemispheres: 6000 total pulses). The other group received 20 Hz rTMS (1500 pulses in each of the two hemispheres) but in two sessions spaced by 80 minutes (also totalling 6000 pulses). They showed that while both groups received 6000 pulses per day, the group receiving two sessions responded faster than the ones receiving only a single session. They claim that additional pulses in accelerated rTMS protocols likely render greater benefits when delivered with optimal intervals of approximately 80 minutes between sessions. Together these studies suggest that simple prolongation of rTMS protocols are likely to be less effective than applying rTMS in multiple sessions separated adequately in time.

However, it must be noted that a recent trajectory analysis of treatment response in two groups receiving either accelerated or non-accelerated rTMS, did not find evidence of either faster or larger antidepressant action [116]. However, Kaster et al. used an accelerated rTMS protocol that was still spread over 4 weeks, as the accelerated protocol only reduced the number of days per week on which stimulation was delivered. This may have resulted in no significant differences in response trajectories of accelerated and non-accelerated rTMS. In addition, variations in inter-session intervals suggests a lack of consensus for an optimal inter-session interval for accelerated rTMS protocols. Hence, a sub-optimal inter-session interval may also have resulted in lack of greater clinical benefits from accelerated rTMS as compared to standard rTMS. Thus, future works comparing accelerated rTMS protocols delivered within a week to the standard rTMS protocol will help further delineate its clinical benefits and optimize the inter-session interval. Nevertheless, the current evidence indicates that while accelerated rTMS protocols are feasible, they may not significantly influence the antidepressant efficacy, but only reduce the burden of treatment on patients.

1.4.3 Resting state fMRI - a newer neuroimaging tool

While RCTs established the efficacy of rTMS in the treatment of depression, the advancements in non-invasive neuroimaging have furthered the understanding of the pathophysiology of psychiatric disorders.

This has led to modelling of these disorders as dysfunctions of large-scale brain networks. Functional magnetic resonance imaging (fMRI) has been a critical aid in visualizing the functioning of a brain *in vivo* with high spatial resolution. It primarily quantifies blood oxygenation level dependent (BOLD) signal changes, which are tightly coupled to hemodynamic response of areas of activation in response to cognitive events and stimuli [117]. With time, a paradigm shift has allowed further comprehension of the organizational architecture of psychiatric disorders with resting state fMRI (rsfMRI). RsfMRI focuses on naturally occurring slow fluctuations in BOLD signals with frequencies lower than 0.1 Hz to evaluate brain regions that preferentially work together along time (in coherence) [118]. It was first demonstrated by Biswal et al. [119] where synchronous fluctuations of the left somatosensory cortex identified its counterparts on the contralateral side. Further work established the existence of multiple large-scale brain networks defined by coherent BOLD activity in distant brain regions. The coherent BOLD fluctuations between distant brain regions define the magnitude of functional connectivity of those regions. Higher synchrony imply higher functional connectivity, a complete lack of synchrony implies a negative functional connectivity i.e. anticorrelation [120]. The latter means that a stronger activation in one region is accompanied by weaker activation in another in contrast to the general average, i.e. the BOLD fluctuations are interlocked with each other out of phase.

The analysis of rsfMRI data involves state-of-the-art preprocessing steps involving slice time correction, head motion correction, regression of motion parameters, white matter, and cerebrospinal fluid signal [121]. When possible, signals corresponding to the cardiac rhythm and the respiratory cycle are handled as nuisance factors and regressed out of the fMRI data [122, 123]. To further boost the signal-to-noise ratio, these steps are followed by spatial smoothing and filtering of the signal to retain only slow neural fluctuations relevant for resting state inside the frequency band of 0.01 − 0.1 Hz [124, 125]. The data can also be transformed into MNI standard space to allow group-based averaging and statistical analysis [117, 126]. The rsfMRI data in standard space can be further analysed using two common methods: seed based and independent component analysis (ICA). By using either of the approaches, one can identify various networks that constitute the functioning of the brain. For the seed-based method, one selects an *a priori* region of interest (ROI) and maps voxels whose time courses have a high positive or negative synchrony (i.e. functional connectivity) to the time course of the ROI [118, 127]. For the ICA, one does not require any *a priori* selection of ROI as the whole-brain information is computed. The ICA algorithm attempts to segregate the rsfMRI signal into mutually exclusive and functionally independent spatial components called large-scale networks [126, 128, 129].

One such prominent rsfMRI network is called the default mode network (DMN), first reported in 2001 by Raichle et al. [130]. This network comprises the medial and lateral temporal lobes, medial prefrontal cortex and medial and lateral parietal cortex [131]. Its main function, while still not completely known, spans the domain of emotional processing, self-referential thoughts and memory recollection [132]. Greicius et al. [133] were first to identify the DMN using rsfMRI data, further confirmed by others [128, 134–136]. As multiple works established the veracity of such large-scale networks, it led to development of network-based models of brain functioning. It states that the brain is comprised of spatially independent networks – composed of distant brain regions whose BOLD fluctuations are more synchronous to each other than to other brain regions. Apart from the DMN, the central executive network (CEN) and the salience network (SN) are other robust and crucial brain networks. The networks activated in response to external stimuli are termed as the task-positive networks due to their association with cognitive tasks performed by the subjects. The DMN, referred to as the task-negative network, shows deactivation in task related manners and is involved in self-directed cognitive processing [137–139]. Together these three networks constitute the triple network model of brain functioning [140, 141]. The CEN is associated with performance of cognitive tasks processing external stimuli [142, 143]. The DMN is involved in more inward, self-referential thought processing and its activation is associated with downplay of the CEN. These networks activate in response to the external or internal stimuli leading to two opposing systems [117]. The third major network, the SN plays the role of functional switching between the CEN and the DMN depending on the task at hand and thus acts as a circuit breaker to commence relevant processing of stimuli as appropriate [140].

1.4.4 Understanding depression – A perspective from large-scale networks

The use of rsfMRI in patients with depression has shown that functional connectivity aberrations in large-scale brain networks underlie depression pathophysiology. The DMN plays an important role in self-directed processes and several rsfMRI studies have reported DMN connectivity aberrations in patients with depression. Greicius et al. [144] acquired rsfMRI from twenty-eight patients with depression and twenty healthy individuals. They found that the functional connectivity of the DMN to the sgACC, thalamus, orbitofrontal cortex, and precuneus is higher in depressed individuals compared to healthy controls. A similarly increased connectivity of the sgACC to the DMN in depressed cohort was also reported by Hamilton et al. [145] by means of Granger Causality analysis (GCA). They reported increased excitatory interactions between the sgACC and medial prefrontal cortex (MPFC). In addition to an inhibitory influence between the hippocampus and DLPFC, they also found a negative influence of sgACC over the posterior cingulate cortex (PCC), DMPFC, and ventral striatum. These findings were further supported by Liston et al. [146] who found an increased functional connectivity within the DMN as well as an increased connectivity of the DMN to the sgACC, ventromedial prefrontal cortex (vmPFC), pregenual anterior cingulate cortex (pgACC), thalamus, and precuneus. Such findings postulate that aberrations seen in interactions of several brain regions influence the symptomatology of depression, rather than separate deficits in brain regions. For example, in a task-based study Sheline et al. [147] found that patients with depression have significantly larger DMN responses to negative pictures, which depicts a failure to regulate DMN activity. A relationship between aberrant functional connectivity and behaviour was also highlighted by Berman et al. [148] who found a significant association between rumination and the functional connectivity of the sgACC and PCC (both are nodes of the DMN). These associations between behaviour and inherent functional connectivity were further strengthened by Zhu et al. [149]. They reported a positive association of rumination with functional connectivity between the DMPFC and temporal pole as well as with the functional connectivity of the lateral temporal cortex and parahippocampal cortex. The differences in the regions between the two studies may stem from differences in how rumination was scored, patient characteristics, study design, and analysis methods. It nevertheless points out that depressive symptoms are mediated not by the abnormality of a single region. Instead, depressive disorders are resultant of aberrations in the interactions of several regions, which function rather coherently as a large-scale network in the absence of pathophysiology.

The CEN is the task positive network with the DLPFC as an important network node. The CEN plays an important role in exercising cognitive control via DLPFC function. Hamilton et al. [77] found that the DLPFC is significantly less reactive to negative stimuli and less active at rest in depressive disorder. They investigated if the two kinds of dysfunction within the DLPFC overlap with each other, i.e. if both dysfunctions arise from the same region of the DLPFC. They projected the coordinates that show reduced activity at rest and those that show under-reactiveness to negative information to a common template. While the reduced activity during rest is localized to the left DLPFC, the under-responsiveness to negative stimuli is attributed to the right DLPFC. Note, that both the regions are still a part of the CEN as defined by Seeley et al. [150], however the differential role attributed to the contralateral DLPFC in Hamilton et al. [77] suggests that left and right DLPFC regions may differentially affect the depression pathophysiology. Additionally, Liston et al. [146] also reported a reduced functional connectivity within the CEN compared to healthy controls. Specifically, they reported reduced functional connectivity of the left DLPFC and several brain regions like the premotor cortex, posterior parietal cortex, and parts of the prefrontal cortex. These findings depict depression pathophysiology as marked by abnormal connectivity between brain regions that together constitute the CEN.

The third important network is the SN comprising of the dorsal Anterior Cingulate Cortex (dACC), anterior insula (AI), and the amygdala [77]. It plays a crucial role in directing attention to relevant environmental stimuli. Past research has shown that these nodes are overactive in response to relevant stimuli (e.g. dACC to sad faces [151], insula to monetary loss [152], and amygdala to anticipation of aversive visual cues [153]). At rest in non-task conditions, Liu et al. [154] analysed the regional homogeneity of the insular region, measuring its local connectedness, in patients with depression. They found the right insula had a lower regional homogeneity compared to healthy subjects indicating a lower level of connectedness in depression pathophysiology. Further supporting this finding, Manoliu et al. [142] reported a decreased functional connectivity between the bilateral AI and the SN indicative of poor within SN connectivity. This reduced right AI-SN connectivity was associated with increased depression severity, further consolidating the link between large-scale network connectivity and depression symptomatology. For the dACC region, one study found a reduced functional connectivity of the dACC and the left angular gyrus, bilateral middle frontal gyrus, and precentral gyrus [155]. Together, the findings from the DMN, CEN, and SN suggest that within large-scale network functional connectivity aberrations underlie and describe depression pathology.

Apart from within network aberrations of functional connectivity, the abnormal interactions between networks also underlay depression pathology. Manoliu et al. [142] reported a decreased connectivity between the subsystems of DMN and the CEN. The connectivity of these DMN and CEN subsystems also correlated positively with the reduced connectivity between right AI-SN. Their results indicate that within network connectivity alterations may be related to (if not causing) between network connectivity aberrations. Such altered network interaction is further reinforced by Goya-Maldonado et al. [156] who reported reduced functional connectivity between the SN and DMN. A meta-analysis also reported similar hypoconnectivity between subsystems of the SN and DMN [157]. They analysed the affective network (AN) and ventral attention network (VAN), referred together as the SN. They found that nodes of AN (parts of the ACC, amygdala, and nucleus accumbens) showed reduced connectivity with MPFC regions, belonging to the DMN. They also found reduced connectivity between nodes of the VAN (parts of ACC, insula etc.) and precuneus and PCC (both nodes of DMN). Considering the role of the SN in switching between DMN and CEN [141], these studies suggest a network dysfunction of depression where reduced functional connectivity within the SN invokes failure to mediate switching between the DMN and CEN. This likely results in an overactive DMN yielding a continued ruminative state, which often characterizes depression [77, 156]. However, Mulders et al. [158] have reported contradictory findings of an increased functional connectivity between the anterior DMN and the SN. This may be a consequence of methodological differences between studies in terms of network definition (analysing subsystems of large-scale networks), and different methods of analysis (ICA versus seed-based network formulation). Methodological differences aside, these reports demonstrate that depression pathology can be mapped and studied as aberrations of complex network interactions between several brain regions. Most notable of such aberrations are observed within and between the DMN, CEN, and SN.

1.4.5 Understanding brain network effects of therapeutic rTMS in depression

RsfMRI has shed light on both between and within large-scale network abnormalities associated with depression pathology. Acquiring rsfMRI before and after rTMS has further facilitated an understanding of large-scale network mechanisms associated with its antidepressant effects. In one such study, Liston et al. [146] combined 10 Hz rTMS with pre- and post-treatment rsfMRI and reported on network mechanisms of antidepressant action of HF-rTMS. After 10 Hz rTMS over the left DLPFC they found the sgACC hyperconnectivity to DMN was reduced post treatment. They also reported that higher functional connectivity of sgACC to DLPFC and bilateral parietal cortex related to higher treatment response, indicating antidepressant effects from rTMS may be mediated by the functional connectivity of these regions. No significant changes in the CEN connectivity were reported after 10 Hz rTMS. This led them to theorize that the mechanism of 10 Hz rTMS in depression treatment likely involves normalizing the DMN connectivity, especially that between sgACC and DMN. Their results are in line with previous works that show reduced metabolic activity and blood flow in sgACC in response to pharmacotherapy (selective serotonin reuptake inhibitor [SSRI] and serotonin-norepinephrine reuptake inhibitor [SNRI]) and ECT, respectively [159]). However, cognitive behaviour therapy (CBT) increases the metabolic activity of sgACC [159]. The opposite results may likely be a result of different mechanisms corresponding to each intervention, but what is important to note is that antidepressant mechanisms likely involve changes in the sgACC regions. This role of sgACC in antidepressant response is further supported by another study using accelerated 20 Hz rTMS protocol. Baeken et al. [160] acquired rsfMRI before and after a complete study protocol (including real and sham rTMS) in patients with depression. They reported that responders were characterized by a greater functional connectivity between sgACC and parts of the DMN (left superior frontal gyrus). They also found an increased functional connectivity of the sgACC and the perigenual prefrontal cortex (pgPFC), a region involved in emotional regulation. Considering the role of pgPFC in emotional regulation, Baeken et al. [160] conjecture that the increased connectivity between sgACC and pgPFC may result in better regulation of affect and might be important for depression symptom alleviation. Their results indicate a different mechanism of sgACC regulation in response to rTMS. This however may be a result of different rTMS protocols used by respective studies. Liston et al. used a 10 Hz rTMS protocol applied over 4 weeks. On the other hand, Baeken et al. used an accelerated 20 Hz rTMS protocol applied within a week. The differences in how the sgACC responds to these protocols likely stems from differences in the rTMS protocols. Furthermore, differences in study design and seed region definition might also have resulted in variable effects reported on sgACC connectivity.

Aside from HF-rTMS, the functional connectivity of sgACC region has also been implicated in antidepressant response to iTBS. In a study using accelerated iTBS (aiTBS), Baeken et al. [161] found an increased sgACC connectivity to the medial orbitofrontal cortex (mOFC) in iTBS responders. They also found the increase in sgACC-mOFC connectivity to correlate with the decrease in feelings of hopelessness. This further suggests that changes in functional connectivity of the sgACC possibly mediate symptomatic alleviation in depression treatment. As reported for other antidepressant treatments [159], the involvement of the sgACC and regions of the DMN is consistently reported in antidepressant response from rTMS modality. This is apt, considering the extent of DMN connectivity aberrations that underlay depression pathophysiology.

Combining rsfMRI and rTMS to delineate the underlying DMN mechanisms is an important step to further understand the network level effects of these treatments and how such effects may translate to symptomatic improvement at the behavioural level. The precise understanding of effects resulting from rTMS will potentially encourage exploration of factors that can predict such effects. A lucid understanding of association between factors that predict rTMS induced network effects and their relationship to symptom improvement will promote pre-emptively choosing a relevant rTMS treatment modality for individual patients. This will support data driven decision making to assess individual determinants (e.g. genetic, physiological, behavioural, neuroimaging etc.) when choosing the most promising treatment modality, in lieu of a one-solution fits all approach to prescribing medical treatment. Hopefully, such an approach will quicken the therapeutic response times and possibly garner larger therapeutic benefits. The establishment of such predictors will be an important step towards a personalized medicine approach [162]. In this context, the rsfMRI – rTMS combination has also shed light on functional connectivity predictors of rTMS antidepressant response, advancing the field of personalized rTMS approaches.

1.4.6   Functional connectivity predictors of therapeutic rTMS in depression

Several studies have reported multiple brain regions and networks whose functional connectivity is associated with a therapeutic response from rTMS treatment of depression. The functional connectivity of the left DLPFC (the frequently employed target for rTMS) and sgACC (consistently reported in rTMS antidepressant mechanism) have received notable attention in this regard. In retrospective and prospective analyses, studies have reported that left DLPFC sites with higher anti-correlation to the sgACC are associated with greater benefits from rTMS treatment [161, 163–166]. This body of work indicates that left DLPFC targets for rTMS can be chosen based on its anti-correlation to sgACC in an effort to improve the therapeutic response in patients. Connectivity of other brain regions, aside from the sgACC and DLPFC connectivity, also have associations with behavioural improvements in response to rTMS treatment e.g. larger functional connectivity of sgACC with DMN [146], stronger negative connectivity of sgACC and DMPFC [160], reduced functional connectivity to insula, amygdala, parahippocampus, and putamen [166], and sgACC-mOFC connectivity. The reduced DLPFC-caudate [167] connectivity, and increased DLPFC-striatum [168] connectivity can also predict antidepressant response of rTMS. Additionally, functional connectivity of regions other than DLPFC or sgACC have also been reported for their predictive value. For example, higher amygdala-MPFC connectivity, and higher PCC – insula connectivity has been found to predict response [169] and non-response [170] to rTMS, respectively. The connectivity of networks has also been found to predict therapeutic improvement in response to rTMS. For example, the aberrant connectivity of hub nodes of DMN and SN could successfully distinguish treatment responders and non-responders [171]. Similarly, the SN connectivity has been noted to predict clinical response to iTBS at 1 month post treatment [172]. These results convey that there are ample functional connectivity predictors of rTMS response. However, it must be mentioned that these predictors have been reported from studies using different rTMS protocols (5 Hz, 10 Hz, 20 Hz, iTBS), applied in either standard or accelerated manner. The methods of analysis to evaluate functional connectivity are also variable between studies, with some using ICA while others using seed-based analysis. This hinders conclusions on universality of the predictors. Further, while such a varied list of predictors can help clinicians choose between different rTMS protocols, it must be noted that a factor predicting response to a certain rTMS protocol does not necessarily imply non-response to alternative rTMS protocols. Hence, multivariate models parsing together multiple functional connectivity predictors will help evaluate and determine the most probable rTMS protocol for maximum therapeutic outcome.

Such multivariate modelling of predictors could also include factors other than those derived from functional connectivity analysis e.g. from the domain of genetics, behaviour, neurophysiology etc. Silverstein et al. [173] and Noda et al. [174] have summarized various factors in domains other than functional connectivity, that are reportedly associated with rTMS treatment response. These factors include the genetic polymorphism of BDNF and serotonin transporter (5-HTT) related genes; measures of intracortical facilitation (ICF), cortical silent period (CSP), short interval intracortical inhibition (SICI) that correlate with symptomatic improvement post rTMS; and EEG markers showing relationship between alpha, beta, gamma, and delta band characteristics and rTMS response. Furthermore, perfusion, blood flow patterns as well as glucose metabolism of diverse brain regions (sgACC, DLPFC, OFC, etc.) also carry predictive value for response to rTMS. Multivariate modelling of all such factors will help delineate which factors or their combinations, contribute to improved likelihood of rTMS response. A detailed understanding of how factors from such different domains together influence response to rTMS will allow tailoring treatment to each patient, presumably reducing response duration and increasing the therapeutic effects. Additionally, it will expand our understanding of the disease itself, precipitating a more accurate diagnosis and more novel therapeutic approaches. However, the utility of such multivariate modelling in clinical practice to prospectively allocate patients to certain treatment conditions remains to be tested. This is likely since acquisition of such predictors often require access to either an MRI facility, genotyping laboratory, EEG equipment, TMS devices etc. Future studies utilizing each of the predictor modalities in prospective clinical trials will help advocate the adaptation of such a precision medicine approach in clinical practice. An alternative is provided with phenotypic traits, which are typically cheaper to acquire and analyse. Personality traits, for example, can be administered and analysed using only a computer and their impact on depression treatment outcome has also been reported in previous works [175, 176].

1.4.7 Personality predictors of therapeutic rTMS in depression

Personality is a complex human trait arising from a confluence of biology, psychology, genetics, and physiology [177]. According to Cloninger's theory, personality is a composition of two sub-parts: the temperament and the character. Temperament is hereditary differences based on skills and habits [177] likely an outcome of regulation from striatum, amygdala, and other parts of the limbic system [178]. The character is shaped by the social upbringing of the individual and is likely influenced more by the encodings of the hippocampal region along with those of the cortex [178]. These two sub-parts are in turn outlined by various dimensions as discussed further.

As per Cloninger's model of personality, the character is described by three dimensions: self-directedness (SD, self-esteem and responsibility), cooperativeness (tendency to cooperate and view others as part of oneself), and self-transcendence (ST, fulfilment and spirituality)[177]. The temperament is described by four dimensions: harm avoidance (HA), reward dependence (RD), novelty seeking (NS), and persistence (P)[179]. The behavioural propensities of individuals on each of these dimensions is likely influenced by neurotransmitter levels. HA, which refers to an individual's tendency towards behavioural inhibition is mediated by GABA and serotonin. RD mediated by noradrenaline and serotonin, refers to individual tendency for continuance of ongoing behaviour dependent on positive outcomes. NS describes an individual's predisposition to seek greater exhilaration from novel stimuli and is mediated by dopamine.

The dimension P, that refers to tenacity and determination of an individual, is manifested by glutamate and serotonin levels [177]. Note that the evidence for such relationships between neurotransmitters and personality dimensions is mixed, and one-to-one relationships between personality traits and neurotransmitters is rather unlikely. Hence, the manifestation of personality can be described as a complex interaction between not just levels of different neurotransmitters but also functional interactions of distinct brain regions. This is evident from studies showing combinations of personality traits (high HA and low NS) are correlated to distinct connectivity patterns in the brain [180] and the HA scores were correlated with the connectivity of AI and rostral ACC (rACC) [181]. Similar findings are also reported when personality is quantified based on a different theory like the five-factor solution of personality, also called "the Big Five". "The Big Five" was originally based on the assumption that relevant aspects of personality could be gleaned from language. The five factors which measure personality include the openness to experience, extraversion, neuroticism, agreeableness, and conscientiousness [182]. By using "the Big Five" data from the Human Connectome Project, Dubois et al. [182] reported that openness to experience and extraversion could be reliably predicted using multivariate modelling of brain functional connectivity. These results establish a relationship between personality and large-scale brain functional connectivity, and caution must hence be taken when translating molecular findings to complex behaviours such as personality or vice versa [183].

Despite the complex manifestation of personality, extant literature suggest links between personality types and propensity to suffer from depression as well as the likelihood of antidepressant response. Previous works have reported links between high HA and depression and anxiety [184], and between low SD and depression [185]. Additionally, Abrams et al. [184] reported HA to be higher in depression cohorts which reduced significantly after 12 weeks of antidepressant treatment. This finding was reproduced by Hruby et al. [177] who found a significant decrease in HA along with ST scores after treatment and an increase in persistence. A similar reduction in HA scores was also reported by Quilty et al. [186] who additionally reported an association between the reduction in HA and symptom reduction. Aside from reports of antidepressants influencing personality measures, studies have also reported on predictive value of personality for the antidepressant response itself. Kampman et al. [175] showed that high HA scores are correlated to poor antidepressant treatment response. This was further confirmed by Balestri et al. [176] who also found higher HA scores associated with non-response and non-remission in depression. In addition, they also found lower SD was similarly associated with non-response and non-remission [176]. Considering this relationship between depression, pharmacological antidepressant response and personality, presumably personality would also be predictive of the rTMS antidepressant response. However, literature on the relationship between rTMS antidepressant response and personality traits is scant relative to reports on relationship between pharmacological treatment and personality. One such study by Baeken et al. [187] delivered 10 Hz rTMS in thirty six depression patients and found that only SD could predict the outcome of the stimulation. Differently from them, Siddiqi et al. [188] applied up to thirty sessions of 10 Hz rTMS over left DLPFC in 19 patients (some also received LF-rTMS at right DLPFC) and reported a correlation between symptom improvement and persistence. The differences in sample size, treatment protocol, and patient selection might have contributed to divergent results between the two studies. Other than this, Kopala-Sibley et al. [189] delivered rTMS over the DLPFC in patients with depression and quantified personality using "the Big Five" inventory. They found that neuroticism, which is closely associated with the HA measure [190], significantly decreased after stimulation, but was not predictive of symptom improvement.

Given the complex interaction between personality traits and antidepressant treatment response, it would be crucial to understand the relationship between personality and basic rTMS effects. To understand how personality traits might influence left DLPFC iTBS response to stress induced from the Trier Social Stress Test (TSST), Pulopulos et al. delivered two sessions of iTBS in healthy individuals [191]. They reported that while just two sessions of iTBS did not influence mood or cortisol secretion, cooperativeness (measured using TCI) was related to decreased cortisol output. Such delineation of relationship between personality and iTBS effects in healthy individuals is important for understanding how individuals differ in their response to neuro-modulatory techniques in relation to personality. This will likely advance a personalized medicine approach as it would help inform the choice of rTMS based on personality measures and the intended neuro-modulatory effects. The study of the relationship between personality traits and rTMS response in patient cohorts will help inform the role of personality in predicting therapeutic response to neuromodulation. In future, this may help clinicians make decisions regarding the course of treatment while considering individual phenotypic information.

1.5 Heart rate variability and depression

Several regions implicated within the pathology of depression are also involved in autonomic regulation [192–194], which likely explains the association between cardiovascular diseases and depression [195, 196]. The heart rate variability (HRV) is an easy to obtain index for gauging the variation in heart beat and the autonomic control of its functioning [197]. HRV measures are broadly categorized into three classes: time domain, frequency domain, and non-linear measures. The frequently reported time domain measures include the RR interval mean (RR interval refers to the time between two R-peaks), standard deviation of normal to normal RR interval (SDNN), and root mean square of successive heartbeat interval differences (RMSSD) [195]. The mean RR interval is simply the mean of the time differences between two consecutive R-peaks (of the QRS curve). The SDNN measures the standard deviation of normal to normal beats. The beats are referred to as normal after ectopic beats and abnormal beats are cleaned out during preprocessing. The RMSSD is obtained by calculating first the successive time differences between heartbeats. The differences are squared and then the square root of the average finally yields the RMSSD. This measure is often cited as an index of the parasympathetic output of the autonomic system, where higher RMSSD indicates higher parasympathetic activity [198]. Calculating the frequency domain measure involves Fourier transformation of the RR interval time series. The power spectrum thus obtained has three commonly reported frequency domain measures: the very low frequency band (0.0033-0.04 Hz; VLF), the low frequency band (0.04-0.15 Hz, LF) and the high frequency band (0.15-0.40 Hz, HF). The sympathetic functioning largely contributes to VLF band rhythm. The sympathetic and parasympathetic system together induce rhythms that contribute to the LF band. The HF band is largely contributed by the parasympathetic system and pertains to heart rate variations in relation to respiration [198]. The LF/HF ratio is thus indicative of the balance between the sympathetic and parasympathetic systems. Nonlinear measures of HRV such as the Poincare plot, entropy, and fluctuation analysis are less often reported. A more detailed review of these nonlinear domain measures can be found in Schaffer and Ginsberg [198].

The parasympathetic and the sympathetic system [196], which comprise the autonomic nervous system (ANS), mediate the fluctuation in heart's functioning. These changes are mediated in response to stressors and cues from the environment, and a successful adaptation to environmental factors would be the ability of the ANS to modulate physiological response [199]. Thus, large HRV is associated with successful regulation of emotion [199]. The neural regions that together control the physiological response are referred as the central autonomic network (CAN). This network comprises the anterior cingulate, insula, amygdala, and hypothalamus in addition to other regions [197]. CAN together with the parasympathetic and sympathetic nerves innervating the heart is also referred to as the frontal vagal pathway [200]. The CAN mediates its influence on the heart via preganglionic sympathetic and parasympathetic neurons. The signal travels through these neurons from the CAN nodes to the sino-atrial node in the heart. The vagus nerve and the stellate ganglia carry these signals innervating the heart. The complex interplay of these inputs from vagus nerve and the stellate ganglia dynamically adjusts the heart rate [197].

The brain regions constituting the CAN largely overlap to those reported in the pathophysiology of depression [142, 146, 201]. The association between cardiovascular diseases and depression [195, 196] raised the question if HRV aberrations are inherent to depression or merely a consequence of cardiovascular diseases. Evidence indicates that HRV aberrations are inherent to depressive disorder, independent of cardiovascular complaints [196, 202–205]. Furthermore, while there is evidence that even medication free patients with depression have reduced HRV [206], it has also been reported that reduced HRV is observed only in medicated but not in non-medicated patients with depression [207]. This indicates a plausible role of pharmacological treatment in further deteriorating HRV. The effects of pharmacological antidepressant treatment on the HRV are mixed. A review by van Zyl et al. [208] found a decline in measures of HRV in response to tricyclic antidepressants (TCAs), but SSRI use was associated with an increase in HRV and a decrease in heart rate (HR) [208]. The effects of TCAs on HRV were further confirmed through a meta-analysis [196]. However, unlike van Zyl et al., the meta-analysis did not find any significant effects of any other classes of medication on HRV, including the SSRIs [196]. Apart from pharmacological antidepressants, other interventions have also been reported to influence the HRV and HR. The effects of CBT on HRV and HR was studied in 30 patients with coronary heart disease (CHD), both with and without (control group) comorbid depression [209]. The study reported a reduced HR and an increased HRV after up to 16 sessions of CBT, but only in patients who had severe depression. Both non-depressed patients who did not receive CBT and mildly depressed patients receiving CBT, did not show any changes in HRV or HR. However, the generalization of their results is limited, since the sample consisted of patients with CHD. Further studies exploring HRV and HR in patients with depression without comorbid heart diseases will help clarify the effects of CBT on autonomic function. Similar effects have been reported for ECT [210], HRV biofeedback (controlled breathing exercises in response to HR changes)[211], and vagus nerve stimulation [212] (stimulation of left vagus nerve for antidepressant effects [213]). However, the generalizability of effectiveness of such interventions for HRV improvement in patients with depression is limited owing to the small sample size and lack of effective control groups in these studies.

Another significant intervention for treatment of depression, non-invasive brain stimulation (NIBS), has been reported to reduce blood pressure, the HR and increase the HRV in healthy adults [214–216]. The evidence for HRV modulation by NIBS in patients with depression is however mixed. Transcranial Direct Current Stimulation (tDCS) of the left DLPFC showed no influence on HRV or HR in a study of 120 depressed patients receiving either placebo, sertraline, tDCS, or a combination of both the treatments [217]. However, 15 Hz rTMS of the left DLPFC increased the HRV and decreased the HR [218, 219]. Similarly, a 3-minute iTBS at the left DLPFC in patients with depression also resulted in decrease of blood pressure and HR and an increase of HRV in the real condition but not in the sham condition [220]. The variable effects of rTMS and tDCS on HRV may be indicative of differences in underlying mechanisms of the two NIBS approaches or merely be reflective of differences in study methodology, patient characteristics, and electrocardiogram (ECG) data analysis. Nonetheless, the existing evidence does suggest modulation of HRV and HR by rTMS applied over the left DLPFC. Considering the role of the sgACC region in rTMS antidepressant response [146, 160, 161], and in autonomic functioning [192, 221, 222], studies have proposed changes in HR as a marker for effective stimulation of the left DLPFC. As the most effective target of stimulation in the left DLPFC are those with greater anticorrelation to sgACC [163–165, 223], such targets will presumably also induce greater HR changes than other nodes of the left DLPFC. Thus, akin to the use of MEP for evaluating the outcome of stimulating motor cortex, HR changes can be used as a prompt estimate for efficient stimulation of the left DLPFC. Such HR changes based rTMS targeting is referred to as neurocardiac guided rTMS (NCG-TMS)[215]. NCG-TMS offers a novel method for precisely delivering rTMS at the left DLPFC, possibly circumventing the need for rsfMRI. However, there is a lack of studies prospectively using NCG-TMS in patients with depression where stimulation is directed at sites based solely on the HR changes it induces. Future studies will further clarify the clinical potential of NCG-TMS by comparing it to personalized rTMS based on other methods such as MRI, EEG, or behaviour. and will delineate the benefits and limitations of this approach to others.

1.6 Understanding brain network effects of rTMS in healthy population

Changes in response to rTMS seem varied and diverse, both in terms of brain connectivity and heart functioning. Despite widespread clinical use, there is a dearth of knowledge on how clinically relevant protocols interact with brain networks in the absence of depression pathophysiology, although some studies in the past have probed the influence of rTMS over brain networks in healthy control populations. Rahnev et al. [224] reported a reduction in resting state functional connectivity between the visual areas of the brain after applying cTBS to the occipital cortex of healthy adults. Another study applied 1 Hz rTMS to one group and cTBS to another group, over the supplementary motor area (SMA) of healthy subjects [225]. Using pre-rTMS and post-rTMS rsfMRI scans, the study found an increase in functional connectivity within the SMA network after 1 Hz rTMS and a decrease after cTBS. In another study, healthy subjects were stimulated with a single twenty-minute session of 20 Hz rTMS at left parietal cortex sites selected based on its functional connectivity to the hippocampus [226]. The study reported associations between hippocampal functional connectivity changes and improvements in associative memory. However, neither of these studies employed clinically used, FDA approved rTMS protocols. Using the FDA approved 3-minute iTBS, Halko et al. [227] stimulated cerebellum sites selected using rsfMRI. They reported that the functional connectivity of the cerebral-DMN network increased after iTBS, and when iTBS was directed at more midline cerebellar targets, then the dorsal attention network (DAN) was affected instead of the DMN. Although using an FDA approved rTMS protocol, this study did not apply the stimulation at left DLPFC, as is standard in clinical practice.

Thus, there is a dearth of studies in healthy subjects exploring the mechanisms and effects of clinically relevant rTMS protocols in the absence of depression pathophysiology. Only one study has attempted to delineate such mechanisms in healthy subjects. They applied 10 Hz rTMS over the left DLPFC to understand its influence on brain networks of healthy participants [228]. They also explored the temporal evolution of these effects. While their results revealed interesting insights into the functioning of 10 Hz rTMS, they did not apply the clinically typical dose (3000 pulses) of the HF-rTMS protocol. Therefore, their findings cannot be taken as a representative of results from a full dose of 10 Hz rTMS in a healthy brain considering dose-dependent effects of rTMS on functional connectivity [32]. A similar study of healthy adults applied a single session of iTBS at a personalized DLPFC site (based on its connectivity to right AI [rAI]) [229]. They found the connectivity of rAI was reduced after the single session stimulation, and the reduction was proportional to the changes in connectivity of the rAI and the DLPFC. They, however, neither evaluated iTBS effect on the DMN nor explored the time lapse of their results. Thus, the temporal dynamics of iTBS induced changes and its effects on DMN remain elusive. Apart from rTMS, several studies have explored the neural underpinnings of pharmacological treatment in healthy adults. For example, effects from both citalopram and mirtazapine have been explored in healthy volunteers. Citalopram results in increased amygdala activation to fearful faces [230], increased recognition of fearful [231] as well as happy faces [232], and an increased attention bias to positive words [231]. Mirtazapine results in the down-regulation of self-referential processing, decreased fronto-insular cortex response to unpleasant events [233, 234], and an increased bias towards positive emotional memory along with a reduced fear processing [235, 236]. Such studies have not only furthered the understanding of neural mechanisms of pharmacotherapy, but also expanded the clinical role of antidepressants for alleviating constructs such as fear of rejection or the inability to experience pleasure in individuals who do not necessarily qualify a diagnosis of depression [237]. The studies of pharmacological treatment in healthy subjects have thus helped in expanding the therapeutic role of such intervention beyond depression.

A nuanced understanding of basic mechanisms of how normally functioning brain networks are engaged by rTMS will promote individualized rTMS treatment by facilitating selection of an rTMS protocol based on its reported effects on large-scale networks [238]. By doing so, it would be more feasible to *"design, control and repair"* [238] a dynamical system, such as the brain, using rTMS. Like pharmacological treatment, such explorations of rTMS in healthy subjects could also motivate novel therapeutic use of rTMS. The time lapse of the effects from such clinical treatment protocols would be of interest too. Understanding how the effects of the stimulation evolve with time can outline how effects from rTMS advance through the large-scale network. This can potentially shed light on causal interactions within and between networks for further evaluation and optimization of treatment protocols in patient cohorts.

1.7 Understanding brain network and cardiac effects of personalized therapeutic rTMS in depression

DLPFC is the preferred stimulation target for therapeutic benefits of rTMS. However, the heuristic methods used to place the TMS coil at the DLPFC have proven to be inadequate in benefiting all patients receiving rTMS. In earlier studies, rTMS was delivered to a point ~5 cm anterior to the motor hotspot (a scalp location which activates the abductor pollicis brevis), which was later reported as inadequate in localizing the DLPFC [239, 240]. Fitzgerald et al. [239] found that stimulating between the 10-20 EEG based markers: F3 and AF3 was a more accurate method of targeting the left DLPFC compared to 5 cm anterior to motor hotspot method. Furthermore, they also highlighted the benefits of neuronavigation by demonstrating higher response rate in patients with depression when rTMS was delivered at structurally localized left DLPFC using neuronavigation than when rTMS was delivered 5 cm anterior to the motor hotspot. The use of neuronavigation to target stimulation at the correct structural landmarks defining the DLPFC was an important step towards personalized rTMS. However, the DLPFC is a large and complex structure [241]. Its role encompasses a variety of tasks ranging from cognition, attention, working memory etc. [242]. Given this extensive role of DLPFC in a variety of brain processes, simply applying stimulation at structurally defined landmarks is likely to be ineffective in modulating more functionally relevant portions of the DLPFC. This is because a structural landmark based DLPFC rTMS undeniably overlooks the functional connectivity variations of the DLPFC [243]. This was further demonstrated by a single pulse TMS study, delivering pulses at left DLPFC in ten healthy participants in a concurrent TMS-fMRI design.

When the coil was placed at the left DLPFC by anatomical means only, the effects from TMS did not propagate to the sgACC in all subjects [244]. Their results suggest that failure to respond to rTMS as an antidepressant, observed in a portion of patients, may be an outcome of anatomical targeting of left DLPFC, which fails to ensure that the effect of TMS propagates to deeper brain regions.

Thus, development, demonstration and deployment of methods that reckon with inter-individual variations in anatomical and functional connectivity architecture are needed for stepping closer to a personalized medicines approach. One such method is to use individual rsfMRI measurements to estimate the target for stimulation. RsfMRI is easy to acquire and does not require the subjects to actively do any tasks. In task-based fMRI the subjects are needed to perform a task from which the target could then be calculated. While it would be generally simple enough for healthy individuals, patients with psychiatric illness might be unable to perform such tasks satisfactorily due to the pathophysiology of the disease. In such cases, target localization dependent on task-based fMRI might fail to yield suitable stimulation points for such individuals. Hence, rsfMRI would presumably be a more appropriate choice for diseased populations. By adopting such a rsfMRI based personalization approach in clinical practice, a) one would ensure that stimulation is delivered only in cases where it is expected to benefit the recipient thus efficiently managing clinical resources, and b) the personalized approach will expectedly maximize the benefit to the patients. RsfMRI based personalized target selection for stimulation, thus offers a new avenue for research and application in clinical settings. However, little is known about the therapeutic mechanisms of such personalized rTMS approaches and how/if they might differ from rTMS applied using current non-personalized clinical practices. As the clinical use of rTMS has become more mainstream, the need for personalized treatment addressing the sources of variability in rTMS response is mounting [238, 245, 246]. Studies exploring personalized neuromodulation in patients with depression remain scant. Siddiqi et al. [247] used an rsfMRI based personalized 10 Hz rTMS approach in patients with traumatic brain injury comorbid with depression and found the approach feasible and safe. Similarly, Cole et al. [248] applied rsfMRI based personalized iTBS in patients with depression and found the approach effective and feasible. These two studies are restricted in their scope of evaluating the feasibility, safety and efficacy of such personalized treatment approaches. However, no studies have yet explored the mechanisms of personalized rTMS. Thus, the question remains if the therapeutic effects from non-personalized rTMS approaches are the same as those resulting from a more focal and personalized approach.

The work presented in the following chapters, aims to introduce a novel rsfMRI based method of personalization, establish its robustness and explore the mechanisms of such personalized rTMS in healthy subjects. The novel method of personalization is further translationally applied in patients with depression and the results are qualitatively compared against a non-personalized approach. Methods of personalization based on rsfMRI data will additionally increase the financial burden on the patients and clinics. Studying the efficacy of current standard approaches against personalized approaches is important to assess the benefits of such methods over the commonly employed heuristics approaches. Studies, such as the one presented here, will help delineate the improvement in treatment outcome from addressing the underlying variability using personalized approaches or a lack thereof. This should further inform clinical practice and promote the additional use of rsfMRI.

2 Scope of the book

While the therapeutic effects of rTMS for psychiatric illness has been established beyond any qualms, the question that looms over the rTMS research community today is how to boost the efficiency of currently used rTMS protocols in clinical practice. This would extend the therapeutic benefits to patients who failed to respond either to pharmacotherapy, CBT or to non-invasive neuromodulation techniques and hence must rely on more demanding or invasive procedures such as ECT and DBS respectively.

2.1 Personalizing rTMS

The localization of stimulation over the brain is a relatively neglected field with few studies prospectively exploring methods of accurately identifying cortical targets. The underlying variability in the anatomical and functional architecture of individual brains makes accurate identification of cortical targets important for inching closer to personalized rTMS approaches. Past works have shown that the most promising path for personalization may be the selection of rTMS cortical targets that have a higher anti-correlation to sgACC, as this can lead to better symptomatic reduction in patients with depression [163–165]. We first begin by briefly summarizing the state of rTMS in depression treatment and highlight the need to personalize rTMS for better therapeutic response (chapter 3). In this context, we introduce a new method to personalize the left DLPFC target for rTMS using a network-based approach in chapter 4. We investigate the robustness and reproducibility of this novel method for further clinical application by calculating the test-retest reliability of the method.

2.2 Default mode network mechanisms of personalized rTMS in healthy population

Even though work has been done to explore the fundamental workings of rTMS by applying it in healthy individuals, a concrete understanding of how these clinically relevant rTMS protocols mediate their effects is still elusive. The principle aim of the first segment of the work is to understand the underlying large-scale network mechanisms of two of the most commonly used rTMS protocols, 10 Hz rTMS (chapter 4) and the iTBS (chapter 5). By means of a double blind (interviewer and subject), sham-controlled, crossover study design we follow the effects of rTMS protocols in healthy subjects over three time-windows spanning a total duration of approximately 50 minutes after stimulation [249]. The DMN is an important network in both the pathophysiology and treatment of depression. Thus, understanding how the DMN interacts with the two rTMS protocols in healthy brains is crucial for understanding the basic antidepressant mechanism of rTMS protocols. While both iTBS and 10 Hz rTMS alleviate depression symptoms, the brain regions and networks they mediate their effects through remains unknown. By using the same study design, we intend to compare the effects precipitated by either of the protocols. This would inform if both protocols differently affect large-scale brain networks even though the underlying therapeutic effect is similar at the behavioural level. If the mechanisms of iTBS protocol is different from 10 Hz rTMS, it could possibly assist the selection of rTMS protocol for clinical benefits based on the pre-existing network abnormalities. If, however the effects are the same across both rTMS protocols, it would help explain non-inferiority of iTBS to 10 Hz rTMS [88]. We also aim to investigate the rTMS induced DMN effects over multiple time windows after stimulation to reveal the underlying temporal dynamics of such effects. Such an understanding will presumably delineate the propagation of rTMS effects through the brain which may help in optimization of the timing of multiple sessions of rTMS. Lastly, considering the benefits of developing phenotypic predictors for rTMS response, we explore how changes induced by rTMS might be related to personality measures. We expect that revealing underlying links between personality and rTMS induced changes will help formulate working hypotheses that could galvanize into future works with rTMS in patient cohorts.

2.3 Default mode network and cardiac changes after personalized iTBS in patients with major depressive depression

No study to date has explored the mechanisms of multiple sessions of personalized rTMS in patients with depression. While some investigations explore the mechanism of anatomically targeted rTMS [146, 161]; how differently (if at all) the functionally (rsfMRI) targeted rTMS affects brain networks is unknown. Importantly, no comparison has been made between such anatomically targeted rTMS and functionally targeted rTMS. It is unknown if the symptoms alleviation from one is greater than the other. To bridge this knowledge rift, in chapter 6, we investigate aiTBS effects over the F3 location (anatomical) and those from aiTBS over rsfMRI based targets using the method detailed in chapter 4. We aimed to compare our personalization approach against the current clinical practice. Since the use of individual T1 scans to target left DLPFC is not clinically common, comparison against F3 based rTMS would be more informative of benefits from our personalized approach over current clinical practice. Finally, we also aim to evaluate our personalization method in the context of using heart rate changes as a biomarker of effective left DLPFC target engagement. We hence acquired ECG data concurrent with brain imaging as well as iTBS. We hypothesize that if heart rate fluctuations are an effective proxy for engaging DLPFC-sgACC connectivity, then using methods to directly target this connectivity (as done in our personalized approach) should produce greater heart rate changes compared to standard F3 stimulation.

3 Transcranial Magnetic and Direct Current Stimulation in the Treatment of Depression: Basic Mechanisms and Challenges of Two Commonly Used Brain Stimulation Methods in Interventional Psychiatry

NIBS equip health care professionals and researchers with tools to treat and diagnose some neurological and psychiatric conditions as well as gain insights into brain functioning. As the volume of published research concerning NIBS has increased exponentially several methodological and scientific caveats have emerged in the field; these include less robust effects as well as contradictory clinical findings. This suggests that the relationship between the methodological aspects and clinical efficacy of rTMS is far from conclusive. The following chapter dwells into a short review of the two commonly used NIBS modalities, namely rTMS and transcranial direct current stimulation (tDCS). We discuss the basics of the rTMS, the ontogeny of rTMS as a treatment for depression, functional connectivity predictors of rTMS antidepressant response, and possible reasons for the reported inconsistencies in NIBS results. We conclude with a tabulation of suggestions for reducing the variability in NIBS response by integrating neuroimaging data with NIBS procedures. We argue that future multi-centre trials will help establish the potency of such approaches and pave the way forward for comprehensive and personalized treatment routines.

My contributions to the manuscripts pertain to the original drafting and revision of the sections concerning rTMS. The article was originally published at the website of S. Karger AG, Basel.

Neuropsychobiology

Review Article

Neuropsychobiology
DOI: 10.1159/000502149

Received: October 8, 2018
Accepted after revision: July 16, 2019
Published online: September 5, 2019

Transcranial Magnetic and Direct Current Stimulation in the Treatment of Depression: Basic Mechanisms and Challenges of Two Commonly Used Brain Stimulation Methods in Interventional Psychiatry

Aditya Singh[a]   Tracy Erwin-Grabner[a]   Roberto Goya-Maldonado[a]
Andrea Antal[b, c]

[a]Laboratory of Systems Neuroscience and Imaging in Psychiatry (SNIP-Lab), Department of Psychiatry and Psychotherapy, University Medical Center Göttingen, Göttingen, Germany; [b]Department of Clinical Neurophysiology, University Medical Center Göttingen, Göttingen, Germany; [c]Institute for Medical Psychology, Medical Faculty, Otto-v.-Guericke University Magdeburg, Magdeburg, Germany

Keywords
Repetitive transcranial magnetic stimulation ·
Transcranial direct current stimulation · Brain stimulation ·
Neuromodulation · Psychiatry · Depression

derstand the mechanisms underlying the effects of magnetic stimulation and low-intensity transcranial electrical stimulation (TES) in order to optimize dosing, methodological designs, and safety aspects.

Introduction

Transcranial neuromodulation driven by repetitive transcranial magnetic stimulation (rTMS) and transcranial direct current stimulation (tDCS) has been found to be a promising noninvasive treatment for a variety of neuropsychiatric conditions [1–8]. Therapeutic utility of these methods has been claimed for psychiatric conditions such as depression, acute mania, bipolar disorder, panic, hallucinations, obsessions/compulsions, schizophrenia, catatonia, posttraumatic stress disorder, and drug cravings; neurologic diseases such as Parkinson's disease, dystonia, tics, stuttering, tinnitus, spasticity, epilepsy; rehabilitation of aphasia or of hand function after stroke; and pain syndromes such as neuropathic pain, visceral pain, or migraines [7–9].

TMS offers potential for higher efficacy and a lower number of adverse effects relative to pharmacotherapy or electroconvulsive therapy. As a result, the importance of this technique and the therapeutic possibilities are exponentially increasing. The most successful example has been the treatment of major depressive disorder (MDD), which resulted in several countries approving its use in clinical settings. The most frequently applied evidence-based treatment approach for MDD is high-frequency (10 Hz) rTMS over the left dorsolateral prefrontal cortex (DLPFC). It has a level A recommendation in the guidelines for good clinical practice, especially for the acute phase of treatment-resistant depression [9].

Similarly, with regard to the application of tDCS for MDD, the current approach is to enhance neural activity in the left DLPFC with anodal stimulation and/or to reduce neural activity in the right DLPFC with cathodal stimulation [10, 11]. tDCS is recognized with a level B of evidence regarding the antidepressant efficacy of anodal tDCS of the left DLPFC in the guidelines for good clinical practice [8].

In spite of the therapeutic success in MDD, there are still several open questions with respect to the general clinical use of these methods in interventional psychiatry. Studies based on knowledge of rTMS effects, usually in the motor area, have led to further research and, consequentially, guidelines for using rTMS protocols in therapeutic practice [9, 12]. However, many unanswered questions persist regarding how these differences in parameters of rTMS protocol might impact treatment efficacy and what are the possible ways forward to reduce such variations for maximal rTMS stimulation benefits. Currently, rTMS and tDCS are usually applied either as a monotherapy or as augmentation to pharmacotherapy and/or psychotherapy. Although the response to these treatment options is better than sham results in randomized controlled trials (RCTs), a large proportion of patients do not respond. It is therefore imperative to seek optimization of the treatment protocol through personalization of brain stimulation. Furthermore, relating to clinical practice, there is good evidence for beneficial antidepressant effects of transcranial stimulation, although to date the appropriate place of this technique in the therapeutic decision tree is still not clearly defined.

In this short review, we briefly outline the basic principles of rTMS and tDCS. Additionally, we describe how rTMS influences functional connectivity (FC) in response to depression treatment and discuss the sources that introduce disparity in the response to stimulation. Finally, we suggest ways to overcome some of these disparities. Applying these is crucial to boost the performance of rTMS in the treatment of depression.

Transcranial Magnetic Stimulation

TMS was first proposed as a method of brain stimulation in 1985 [13]. It facilitates not only relatively focal stimulation of cortical targets, but also more diffuse stimulation of larger brain volumes, thus penetrating deeper brain regions than tDCS can [14]. The basic principle behind TMS remains the same across protocols and works on Faraday's principle, i.e., a changing electric field induces a changing magnetic field of a few tesla, which in turn induces a perpendicular electric current in conductors in the near vicinity, i.e., a population of neurons [15]. There are several adjustable parameters when designing TMS protocols, such as the number of pulses, frequency, train length, and intertrain interval (ITI) [16]. The research has explored TMS protocols with variations of these parameters mostly on motor-evoked potentials (MEPs) and resting motor threshold (RMT) [17].

While single pulses can evoke MEPs that result in muscle responses when targeted at the primary motor cortex (M1), they are usually not enough to induce longer-lasting effects. Hence, the single pulses of TMS are applied in succession, in prescribed repetitive patterns that allow sustained after effects from the stimulation [18], called rTMS. Therefore, the frequency at which the pulses are delivered and the duration for which they are applied [19] contribute to the variations seen in rTMS protocols that impact the after effects observed. Frequencies of rTMS delivery ≤1 Hz are called low-frequency stimulation and induce inhibitory effects; high-frequency stimulation is ≥5 Hz, and induces excitatory effects in brain [17, 20]. Another important factor that can impact the effects of rTMS includes the intensity relative to the RMT at which the stimulation is delivered. Previous work has shown that if high-frequency rTMS is delivered at intensities lower than the RMT it decreases cortical excitability, while if delivered above the RMT it causes an increase in cortical excitability [21]. A recent study also showed that having adequate ITI during high-frequency rTMS may be essential to its efficacy, as the appropriate ITI can prevent the conduction failure of neurons and thus allow rTMS effects to carry through [22]. It has been proposed that the length of ITI may play a role in rTMS protocol efficacy and should be optimized for the disorder or a desired outcome.

Combining rTMS with brain imaging has increased the understanding of rTMS effects and created the potential to dissect anatomical and functional cross-talk between different brain regions [16]. This ability to modulate brain activity and FC using rTMS has led to its use in clinical settings for therapeutic benefits [23], where it is employed to compensate for processes that are disturbed in psychiatric illnesses. For example, in MDD, rTMS over the DLPFC is used to compensate for deficient cortical excitability and FC [24].

Past research using positron emission tomography (PET) has shown left prefrontal glucose hypometabolism in patients with depression [25]. The use of excitatory rTMS as a treatment stemmed from the expectation of correcting for this hypoactivity in the frontal regions, as several groups had shown that the use of rTMS resulted in an increased blood flow in the prefrontal regions [26, 27] and increased activity under the TMS coil [27–29]. Studies have now gathered evidence for the clinical efficacy of high-frequency (10 Hz) rTMS [30, 31] to the left DLPFC, low-frequency (1 Hz) [32] to the right DLPFC, and bilateral rTMS treatment to both the left and the right DLPFC [33]. More recently, a large study established the noninferiority of intermittent theta-burst stimulation (iTBS) [34] over rTMS protocol. It showed that iTBS stimulation did not differ from 10-Hz rTMS treatment in terms of dropout rates, expected side effects, safety, or tolerability as well as clinical benefits [35]. Thus, it has been established that rTMS (left high-frequency, right low-frequency, and bilateral) and iTBS are effective and reliable treatment options for depression, albeit with different degrees of response.

While PET studies have made a case for rTMS in the treatment of depression, other neuroimaging studies using simple and noninvasive techniques, such as resting-state functional magnetic resonance imaging (rsfMRI), have contributed important knowledge about FC aberrations and changes in depression cohorts compared to healthy populations. Studies have reported that depression patients consistently have dysfunctions associated with the default mode network (DMN), central executive network (CEN), and salience network (SN) [36–38]. Several studies have shed light on the importance of the subgenual anterior cingulate cortex (sgACC), an important node within the DMN, and its negative correlation to the stimulation site at the DLPFC for a better therapeutic response to rTMS [39–41]. Such results from studies using rsfMRI in subjects with depression have allowed for building a FC-based model of the disease. Utilizing these FC-based models, it is possible to stimulate such networks to better understand the antidepressant mechanism of rTMS. For example, Liston et al. [37] reported on the effects of 10-Hz rTMS on the functional networks in depressed subjects: higher sgACC-DMN connectivity in these subjects was decreased post-rTMS treatment, suggesting that rTMS acts by influencing sgACC-to-DMN connectivity. This is in line with the general implication of sgACC in depression treatment, where normalizing sgACC hyperactivity is associated with an antidepressant response [42, 43]. Another study [44] using iTBS for antidepressant treatment showed higher sgACC connectivity to the DLPFC and precuneus in individuals with depression, which normalized in response to rTMS treatment.

Numerous FC features have been employed to predict the response to rTMS and/or iTBS in MDD. For example, Baeken et al. [44] reported that positive connectivity between the sgACC and medial orbitofrontal cortex at baseline could differentiate responders from non-responders to accelerated iTBS. Another study reported that the higher FC within DMN and SN characterized responders for both iTBS and 10 Hz rTMS patient groups [45]. In the case of 5-Hz rTMS, there are reports of more negative pretreatment FC between sgACC and DMN predicting clinical response [46]. For 10-Hz rTMS, the connectivity of DLPFC to several brain regions, namely, reduced FC to the left caudate [47], higher FC to the striatum [48], and greater negative FC to the sgACC [39], has been shown to predict treatment response, highlighting the importance of not only DLPFC but also its FC to distant brain regions in the rTMS treatment mechanisms of MDD.

As is clear from above, there seems to be a wide variety of markers predicting rTMS and iTBS responses and symptoms improvement. In future studies, it would thus be important to replicate the predictive values and investigate combinations of markers that increase specificity and sensitivity or even promote the use of either rTMS or iTBS. Such future development of predictive features, possibly with multivariate data analysis, would allow for greater efficiency of rTMS in clinical settings, thereby benefitting the health care system and providing patient satisfaction by preventing treatments being undertaken that are not expected to work.

While rTMS has been established as an effective treatment for MDD, the percentage of patients that respond to this treatment remains low, varying from 40 to 50% [49]. This likely stems from interindividual variability in the response to rTMS. Some of the factors that contribute to these differences include age, gender, variations in

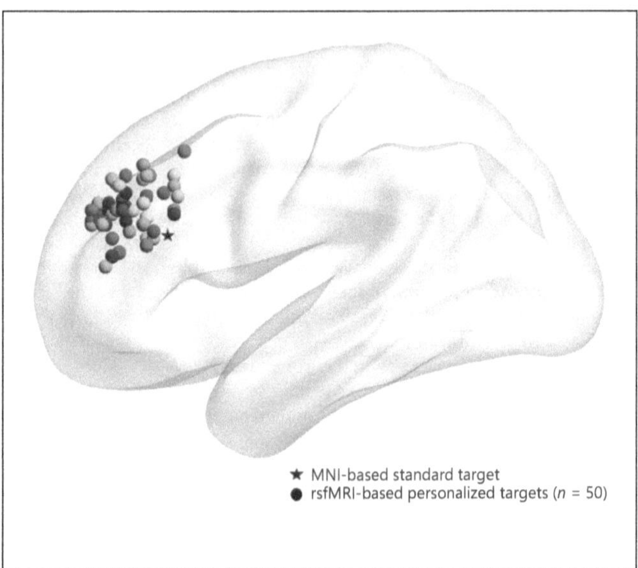

Fig. 1. The standard MNI (anatomical) co-ordinate-based left DLPFC (denoted with a filled star) versus targets selected when employing individual rsfMRI data (denoted with filled circles).

individual cortical excitability, neurophysiological traits, white-matter connectivity, anatomical and functional variability, and genetic polymorphisms [16].

Although age has an influence on the effect of single- or paired-pulse techniques [50], evidence of rTMS being influenced by age remains inconclusive [16]. MEP variability, however, has been shown to be higher in females than in males [51]. White-matter connectivity also influences how the local effects of rTMS spread across networks, so a variation in white-matter pathways would contribute to differences in rTMS effects [52–54]. Similarly, as rTMS is known to cause its effects via tissue reorganization and plasticity (inducing a long-term potentiation/depression-like phenomenon), the genes involved in regulating such events can influence the outcome of rTMS [55, 56].

Other factors such as scalp-to-cortex distance can also influence the outcome, because the magnetic field induced by rTMS decays as a function of distance [27]. Hence, variations in scalp-to-cortex distance between subjects result in nonequivalent amounts of current being induced in the same cortical region when the same threshold, relative to RMT, is used for each individual. In this sense, simple corrective measures to calculate the re-

quired compensatory increase or decrease in rTMS intensity based on scalp-to-cortex distance have been suggested [57].

Apart from the reasons mentioned above, the target site for stimulation is also important to note. In the case of rTMS for the treatment of depression, the common practice of delivering stimulation at 5 cm anterior to the motor cortex has been shown to be ineffective at reaching the desired DLPFC target [58]. Hence, it has been suggested that EEG-based landmarks and structural or functional MRI to guide the rTMS coil to the DLPFC using online neuronavigation systems be used. Figure 1 illustrates standard versus personalized sites for stimulation. However, an anatomically based landmark for targeting rTMS often fails to account for differences in the functional architecture of the brain from individual to individual [59–61]. Thus, using anatomical landmarks can provide a general location of various brain regions, but might still grossly miss the functionally important regions in terms of either network connectivity or functional activity.

We therefore suggest directing rTMS at cortical targets using functional activity-related information. This can be task-based fMRI, which uses a task that engages the region of interest at which the stimulation is then aimed. The idea of engaging the aberrant targeted region arises from the knowledge that the effects of TMS or rTMS are very much dependent upon the activity occurring in the region being targeted [62, 63]. Previous studies [64] have shown that TMS pulses or bursts usually activate the subpopulations of neurons, which are exhibiting the lowest levels of excitability. This is the state-dependency phenomenon of neurostimulation, where the current state of a subject's brain influences the outcome of the stimulation. Therefore, by engaging particular regions of the brain, researchers can probably influence the anticipated impact of rTMS.

Similarly, it can be based on rsfMRI, where network-based FC is used to identify the important nodes of a network that are targeted using rTMS. As described in studies by Weigand et al. [39] and Fox et al. [40, 41] our research group utilized rsfMRI to identify left DLPFC targets based on the connectivity to the sgACC [65]. We selected rTMS target nodes in the left DLPFC that had a higher negative connectivity to the sgACC as it has been suggested that this has a better clinical outcome. Our study forms the basis that it is in fact practical to use individual rsfMRI to target rTMS, hence opening up new avenues for personalization of rTMS treatment.

However, fMRI sessions are currently not a standard part of MDD treatment and care, and the suggestion of personalized targeting of the site of stimulation based on FC data is a preliminary one that requires further validation. This fMRI-based decision could prove important for the 50% of patients who are refractory to therapy and so do not benefit from rTMS [49]. More RCTs exploring the benefits of FC-based versus anatomically based targeting (or 10/20 EEG-system based) are needed; this would encourage the clinical integration of fMRI data for use in rTMS therapy. If the efficacy of such an approach is established from multicenter RCTs, we believe the benefits will outweigh the costs associated with fMRI and would have a positive impact on tertiary health care of patients with MDD.

Transcranial Direct Current Stimulation

Low-intensity transcranial electrical stimulation (TES) methods encompass the external application of electrical current to the brain using at least two electrodes. The externally applied current modulates the spontaneous firing rates of neurons by de- or hyperpolarizing resting membrane potentials (as initially observed in the case of tDCS >50 years ago [66, 67]), thereby changing cortical excitability and activity. Evaluating the functional and behavioral consequences of this method, low-intensity tDCS is particularly appropriate for gaining a further insight into the causative functional role of a given brain area and in functionally connected brain networks, e.g., how brain processes arise and could be changed. The magnitude and direction of the induced after effects are highly dependent on the duration and intensity of the stimulation as well as electrode size and montage [68]. Originally, the tDCS effect was estimated by measuring the amplitude of the MEPs induced by single-pulse TMS [69–71]. Several studies conducted on the M1 showed that the MEP size increased after approximately 10 min of anodal stimulation and decreased after cathodal stimulation [review 71] and remained in this state for up to 60 min after the end of the stimulation. Because this "long-lasting" effect of tDCS is thought to be related to neuroplastic changes in the brain, many studies have addressed the issue of its impact on visual perception and cognitive functions, including motor-learning, working memory, and semantic and episodic memory in healthy subjects as well as its therapeutic applications in neurologic and neuropsychiatric diseases [e.g., 72–79].

In the last few years, studies have been published that question the efficacy of tDCS. High between- and within-group variability and even individual variability were observed, and several of the studies could not be replicated by other investigators. The reason for the relatively high variability is far from being understood [80–85]. Many researchers and clinical practitioners concentrate on the manipulation of four adjustable tDCS parameters: current intensity, stimulation duration, electrode size, and electrode position. These technical variables are easily regulated and play a large role in tDCS effects. However, even these controllable parameters are not always discussed in any cogent manner with regard to increasing efficacy, understanding the mechanisms, or decreasing the variability.

The traditional description of tDCS effects is related to changes in "cortical excitability," i.e., it is thought that the anode that is positively charged enhances the excitability of the underlying cortex while a negatively charged cathode suppresses the excitability of the targeted cortical area [69]. Generally, there is evidence at the neuronal level that anodal tDCS hyperpolarizes the membrane potential in the apical dendritic regions and depolarizes it in the somatic region, whereas the cathodal electrode has an opposite effect [86–90]. Furthermore, besides cell morphology, the extent to which neurons are affected by tDCS (and how) also depends on the orientation of the cells with regard to the induced electric field.

TDCS is typically applied at 1–2 mA (maximum 4 mA), but the electric field in the brain is reduced due to shunting effects (of skin and scalp); it is therefore estimated to maximally reach 0.4–0.8 mV/mm when 1.0 mA is applied externally [91, 92]. While early studies observed that the magnitude and length of the induced after effects (at least after M1 stimulation) increased with higher current intensities and longer stimulation duration [70, 93], later studies reported that doubling the intensity of tDCS led to an opposite effect after cathodal stimulation [94] and increasing the duration of anodal tDCS to 26 min led to MEP decreases [95]. With regard to other cortical areas, the effects of different stimulation intensities and durations have not yet been systematically investigated.

The magnitude and duration of the after effects also depend on the functional state of the brain, i.e., whether the stimulation is given during rest or before/together with a motor or cognitive task [96]. When tDCS has no effect during tasks, it can be speculated that the effect is perhaps too weak to manifest when the activity is being performed [97].

The size of the electrodes and their montage are highly relevant for the efficacy of the stimulation. Conventionally, tDCS involves two electrodes placed on the scalp. Typical electrode sizes range between 4 and 35 cm². A first limitation here derives from the wide electric field induced in the cortex by such large electrodes [86, 91, 98, 99]. The consequence is poor focusing, which can make the interpretation of the results difficult because of the impossibility of precisely locating the structures affected by tDCS. It should also be considered that almost all previous studies targeted single brain regions with low-intensity TES to modulate brain function. However, brain regions interact with each other through networks; the stimulation of a single brain area may thus influence and/ or be influenced by other regions and networks. Because of this complexity, the type of stimulation that was originally seen as "excitatory" (anodal tDCS) might not always increase "cortical excitability" and vice versa. A better description of the tDCS effect in the future might be that it modifies the "excitability-inhibitory balance" in the stimulated and related cortical areas. With regard to the parallel stimulation of multiple regions of a network [100], high-definition multielectrode tDCS arrays, of up to 32 electrodes (HD tDCS), have recently become available [101, 102].

Regarding the use of tDCS for the treatment of depression, the current approach is similar to that described above for rTMS. Most of the clinical trials have concentrated on enhancing the neural activity in the left DLPFC with anodal stimulation and/or reducing the neural activity in the right DLPFC with cathodal stimulation [11, 103–108]. Computer modeling and neuroimaging (fMRI) tDCS studies suggest that, in fact, the stimulation also largely affects deeper brain structures, such as the sgACC, amygdala, and hippocampus [86, 99, 109, 110].

The antidepressant effect of anodal tDCS on the left DLPFC was investigated in at least 15 RCTs [8], as well as in several case reports and open-labeled studies. Unfortunately, most of the RCTs investigating the beneficial effects of tDCS had large patient sample variability (related to the diagnosis of drug-resistant, unipolar, or bipolar depression) and different goals (comparing different stimulation protocols, add-on treatment [pharmacotherapy], or long-term treatment), so no solid conclusions can be made based on these data. One of the critical points is that the concomitant administration of antidepressant medication, benzodiazepines, and antiepileptics can influence tDCS-mediated effects on brain excitability and may indeed have increased variability and reduced the therapeutic impact in these studies. On the other hand, in a recent clinical trial, the combination of anodal tDCS with sertraline (50 mg/day) was superior to each treatment applied alone and to placebo, suggesting an additive interaction of tDCS and antidepressant medication (the SELECT-TDCS trial [111–113]). Previous studies implied that the outcome of tDCS in healthy subjects may be mediated by pharmacological modulation of noradrenergic serotonergic pathways. Therefore, serotonergic enhancement might increase the neuroplastic effects of anodal tDCS, thus resulting in synergistic effects [114, 115].

A precise understanding of the differences between responders and nonresponders may help in the identification of patients responsive to tDCS at the beginning of the therapy. In a recent trial (Escitalopram vs. Electrical Current Therapy for Treating Depression Clinical Study [ELECT-TDCS]), it was observed that the plasma levels of nerve growth factor predicted early depression improvement due to tDCS; this should be explored in further clinical studies [106].

Recent clinical guidelines recommend the following when using tDCS for treating depression: anodal tDCS of the left DLPFC (with right orbitofrontal cathode) delivered for at least 10 days (stimulation intensity: 2 mA and duration: 20–30 min) in medicated or drug-free patients with MDD. There is also an appropriate amount of evi-

Table 1. A summary of possible approaches for reducing variability in the rTMS and tDCS response in patients with depression

First author [Ref.], year	Suggested method of reducing interindividual variability
McClintock [118], 2018	Consensus recommendation by experts in the field of rTMS on effective and safe use of rTMS for depression treatment. The expert panel suggests that MRI-guided coil positioning achieves greater precision; however, evidence of greater efficacy is still limited. In the absence of a neuronavigation system or MRI, coil positioning at F3 based on the international 10–20 system is preferred for clinical use.
Singh [65], 2019	Proof-of-concept study showing the use of rsfMRI connectivity to identify relevant networks covering the left DLPFC and sgACC and identifying functionally relevant nodes within the left DLPFC region that also have a negative correlation to the sgACC
Siddiqi [119], 2019 Siddiqi [120], 2019	Using rsfMRI connectivity-based dorsal attention network and DMN to identify left DLPFC targets based on maximal negative connectivity to the sgACC
Luber [121], 2017 Neacsiu [122], 2018	A case for linking rTMS to the underlying neurocircuitry in depression disorder and using this model to individualize the rTMS targets based on task-based fMRI
Sathappan [123], 2019	A review on how concurrent or sequential use of cognitive interventions can allow controlling the state of the neural network being targeted, facilitating a more efficient and optimized response to rTMS
Stokes [57], 2005	Using T1-weighted images to scale the motor threshold to avoid over- or understimulation
Nord [124], 2019	Anodal tDCS is significantly superior to sham in individuals with high left prefrontal cortex BOLD activation at baseline during the performance of an n-back task (86% accuracy in predicting clinical response using this measure)
Filmer [125], 2019	Pre-tDCS baseline measures of GABA and glutamate, acquired using MRS, correlated with the extent to which stimulation modulated behavior
Bajbouj [126], 2018	CBT combined with active bifrontal tDCS performed during psychotherapy sessions (anode over F3 and cathode over F4, current intensity 1–2 mA, applied 10 min after starting CBT and lasting 30 min) increased the efficacy of the stimulation and resulted in a stronger antidepressive effect
Lefaucheur [7], 2016	Concomitant administration of benzodiazepines or antiepileptics may influence the tDCS-mediated effect on cortical excitability and reduce its therapeutic impact in patients with depression
Brunoni [127], 2013	Concomitant administration of anodal tDCS over the left DLPFC with sertraline hydrochloride (50 mg/day) is superior to each treatment on its own or placebo, suggesting an additive interaction of tDCS and antidepressant pharmacotherapy
Lefaucheur [7], 2016	Good care must be taken to ensure that subjects cannot differentiate between stimulation and control conditions

Of note, these suggestions may have the potential to be applied to other psychiatric disorders. Based on the underlying neural circuitry implicated in the psychopathology, neuroimaging methods can be adopted to more precisely target relevant brain regions/networks.

dence to make a level B recommendation concerning the absence of efficacy using the same tDCS protocol in patients with drug-resistant depression.

Conclusion

rTMS and tDCS have been widely explored for the treatment of depression and their efficacy, safety, and tolerability have been established. By combining these data with rsfMRI information, a developed FC-based model indicates that MDD is associated with aberrant DMN, CEN, and SN connectivity, possibly due to hyperactivity of the sgACC. rTMS is believed to normalize these functional networks, which is reportedly in line with reduced sgACC activity in response to antidepressant treatment. Despite the current understanding, the effectiveness remains low for the treatment of MDD, partly stemming from age, gender, and individual genomic, anatomical, and functional variability. Of these, age and gender vary in the real-world patient population; an effort should thus be made to increase stimulation efficiency independent of age and gender. Table 1 summarizes suggestions to reduce variability in the response to standard rTMS by means of integrating individual neuroimaging data. As mentioned earlier, these recommendations are still in need of rigorous scientific testing. Multicenter RCTs will establish the utility of these approaches, for the treatment of depression and other psychiatric disorders.

Similar to rTMS, one of the critical points in tDCS is to increase the efficacy of stimulation. Early studies used a maximum of 20-min-long sessions of 1 mA anodal tDCS over the left DLPFC and a cathode placed over the right supraorbital region. Current trials support the use of a longer stimulation duration (30 min) and a higher intensity (2 mA), with the cathode placed over the right DLPFC. However, it must still be demonstrated whether increasing the duration and intensity of stimulation automatically leads to improved therapeutic efficacy in MDD. With regard to the long-term effects of tDCS, the number of studies is very limited. In a small study (on 11 patients) who completed a 3-month follow-up after a 10-day tDCS protocol, it was found that 45% of the patients were still responders at the latest time point [116]. Nevertheless, a higher relapse rate has been reported when the repetition of the sessions was reduced from weekly to biweekly [111, 117].

Thus, we believe that attempting to reduce variability in factors such as target location and stimulation protocols, which contribute to the differences in response to brain stimulation, is the way forward for increasing the clinical effectiveness of transcranial stimulation. Furthermore, continued searches for predictive markers of brain stimulation response, molecular/genetic markers, and rs-fMRI-based FC markers for both tDCS and rTMS will likely augment the effective use of brain stimulation. Adequately sensitive and specific markers will allow for more efficient use of both clinical resources, including better utilization of patients' time and the overall impact on health, thus benefiting the economy in the larger run.

References

1 Aleman A, Sommer IE, Kahn RS. Efficacy of slow repetitive transcranial magnetic stimulation in the treatment of resistant auditory hallucinations in schizophrenia: a meta-analysis. J Clin Psychiatry. 2007 Mar;68(3):416–21.

2 Devlin JT, Watkins KE. Stimulating language: insights from TMS. Brain. 2007 Mar;130(Pt 3):610–22.

3 Fregni F, Pascual-Leone A. Technology insight: noninvasive brain stimulation in neurology-perspectives on the therapeutic potential of rTMS and tDCS. Nat Clin Pract Neurol. 2007 Jul;3(7):383–93.

4 George MS, Nahas Z, Borckardt JJ, Anderson B, Foust MJ, Burns C, et al. Brain stimulation for the treatment of psychiatric disorders. Curr Opin Psychiatry. 2007 May;20(3):250–4.

5 Levkovitz Y, Isserles M, Padberg F, Lisanby SH, Bystritsky A, Xia G, et al. Efficacy and safety of deep transcranial magnetic stimulation for major depression: a prospective multicenter randomized controlled trial. World Psychiatry. 2015 Feb;14(1):64–73.

6 Neef NE, Hoang TN, Neef A, Paulus W, Sommer M. Speech dynamics are coded in the left motor cortex in fluent speakers but not in adults who stutter. Brain. 2015 Mar;138(Pt 3):712–25.

7 Lefaucheur JP. A comprehensive database of published tDCS clinical trials (2005-2016). Neurophysiol Clin. 2016 Dec;46(6):319–98.

8 Lefaucheur JP, Antal A, Ayache SS, Benninger DH, Brunelin J, Cogiamanian F, et al. Evidence-based guidelines on the therapeutic use of transcranial direct current stimulation (tDCS). Clin Neurophysiol. 2017 Jan;128(1):56–92.

9 Lefaucheur JP, André-Obadia N, Antal A, Ayache SS, Baeken C, Benninger DH, et al. Evidence-based guidelines on the therapeutic use of repetitive transcranial magnetic stimulation (rTMS). Clin Neurophysiol. 2014 Nov;125(11):2150–206.

10 Brunoni AR, Valiengo L, Baccaro A, Zanao TA, de Oliveira JF, Vieira GP, et al. Sertraline vs. ELectrical Current Therapy for Treating Depression Clinical Trial—SELECT TDCS: design, rationale and objectives. Contemp Clin Trials. 2011 Jan;32(1):90–8.

11 Dondé C, Amad A, Nieto I, Brunoni AR, Neufeld NH, Bellivier F, et al. Transcranial direct-current stimulation (tDCS) for bipolar depression: A systematic review and meta-analysis. Prog Neuropsychopharmacol Biol Psychiatry. 2017 Aug;78:123–31.

12 Rossi S, Hallett M, Rossini PM, Pascual-Leone A, Avanzini G, Bestmann S, et al.; Safety of TMS Consensus Group. Safety, ethical considerations, and application guidelines for the use of transcranial magnetic stimulation in clinical practice and research. Clin Neurophysiol. 2009 Dec;120(12):2008–39.

13 Barker AT, Jalinous R, Freeston IL. Non-invasive magnetic stimulation of human motor cortex. Lancet. 1985 May;1(8437):1106–7.

14 Miranda PC. Physics of effects of transcranial brain stimulation. Handb Clin Neurol. 2013;116:353–66.

15 Wagner T, Valero-Cabre A, Pascual-Leone A. Noninvasive human brain stimulation. Annu Rev Biomed Eng. 2007;9(1):527–65.

16 Valero-Cabré A, Amengual JL, Stengel C, Pascual-Leone A, Coubard OA. Transcranial magnetic stimulation in basic and clinical neuroscience: A comprehensive review of fundamental principles and novel insights. Neurosci Biobehav Rev. 2017 Dec;83(October):381–404.

17 Fitzgerald PB, Fountain S, Daskalakis ZJ. A comprehensive review of the effects of rTMS on motor cortical excitability and inhibition. Clin Neurophysiol. 2006 Dec;117(12):2584–96.

18 Mutz J, Edgcumbe DR, Brunoni AR, Fu CH. Efficacy and acceptability of non-invasive brain stimulation for the treatment of adult unipolar and bipolar depression: A systematic review and meta-analysis of randomised sham-controlled trials. Neurosci Biobehav Rev. 2018 Sep;92:291–303.

19 Simonetta-Moreau M. Non-invasive brain stimulation (NIBS) and motor recovery after stroke. Ann Phys Rehabil Med. 2014 Nov;57(8):530–42.

20 McGirr A, Berlim MT. Clinical Usefulness of Therapeutic Neuromodulation for Major Depression: A Systematic Meta-Review of Recent Meta-Analyses. Psychiatr Clin North Am. 2018 Sep;41(3):485–503.

21 Modugno N, Nakamura Y, MacKinnon CD, Filipovic SR, Bestmann S, Berardelli A, et al. Motor cortex excitability following short trains of repetitive magnetic stimuli. Exp Brain Res. 2001 Oct;140(4):453–9.

22 Halawa I, Goldental A, Shirota Y, Kanter I, Paulus W. Less might be more: conduction failure as a factor possibly limiting the efficacy of higher frequencies in rTMS protocols. Front Neurosci. 2018 May;12(MAY):358.

23 Klomjai W, Katz R, Lackmy-Vallée A. Basic principles of transcranial magnetic stimulation (TMS) and repetitive TMS (rTMS). Ann Phys Rehabil Med. 2015 Sep;58(4):208–13.

24 Philip NS, Barredo J, Aiken E, Carpenter LL. Neuroimaging Mechanisms of Therapeutic Transcranial Magnetic Stimulation for Major Depressive Disorder. Biol Psychiatry Cogn Neurosci Neuroimaging. 2018 Mar;3(3):211–22.

25 Martinot JL, Hardy P, Feline A, Huret JD, Mazoyer B, Attar-Levy D, et al. Left prefrontal glucose hypometabolism in the depressed state: a confirmation. Am J Psychiatry. 1990 Oct;147(10):1313–7.

26 Speer AM, Kimbrell TA, Wassermann EM, D Repella J, Willis MW, Herscovitch P, et al. Opposite effects of high and low frequency rTMS on regional brain activity in depressed patients. Biol Psychiatry. 2000 Dec;48(12):1133–41.

27 Nahas Z, Teneback CC, Kozel A, Speer AM, DeBrux C, Molloy M, et al. Brain effects of TMS delivered over prefrontal cortex in depressed adults: role of stimulation frequency and coil-cortex distance. J Neuropsychiatry Clin Neurosci. 2001;13(4):459–70.

28 Teneback CC, Speer AM, Stallings LE, Pharm D, Spicer KM, Ph D, et al. Changes in prefrontal cortex and paralimbic activity in depression following two weeks of daily left prefrontal TMS. J Neuropsychiatry Clin Neurosci. 1999;11(4):426–35.

29 Loo CK, Sachdev PS, Haindl W, Wen W, Mitchell PB, Croker VM, et al. High (15 Hz) and low (1 Hz) frequency transcranial magnetic stimulation have different acute effects on regional cerebral blood flow in depressed patients. Psychol Med. 2003 Aug;33(6):997–1006.

30 George MS, Lisanby SH, Avery D, McDonald WM, Durkalski V, Pavlicova M, et al. Daily left prefrontal transcranial magnetic stimulation therapy for major depressive disorder: a sham-controlled randomized trial. Arch Gen Psychiatry. 2010 May;67(5):507–16.

31 O'Reardon JP, Solvason HB, Janicak PG, Sampson S, Isenberg KE, Nahas Z, et al. Efficacy and safety of transcranial magnetic stimulation in the acute treatment of major depression: a multisite randomized controlled trial. Biol Psychiatry. 2007 Dec;62(11):1208–16.

32 Berlim MT, Broadbent HJ, Van den Eynde F. Blinding integrity in randomized sham-controlled trials of repetitive transcranial magnetic stimulation for major depression: a systematic review and meta-analysis. Int J Neuropsychopharmacol. 2013 Jun;16(5):1173–81.

33 Zhang YQ, Zhu D, Zhou XY, Liu YY, Qin B, Ren GP, et al. Bilateral repetitive transcranial magnetic stimulation for treatment-resistant depression: a systematic review and meta-analysis of randomized controlled trials. Braz J Med Biol Res. 2015 Mar;48(3):198–206.

34 Huang YZ, Edwards MJ, Rounis E, Bhatia KP, Rothwell JC. Theta burst stimulation of the human motor cortex. Neuron. 2005 Jan;45(2):201–6.

35 Blumberger DM, Vila-Rodriguez F, Thorpe KE, Feffer K, Noda Y, Giacobbe P, et al. Effectiveness of theta burst versus high-frequency repetitive transcranial magnetic stimulation in patients with depression (THREE-D): a randomised non-inferiority trial. Lancet. 2018 Apr;391(10131):1683–92.

36 Cooney RE, Joormann J, Eugène F, Dennis EL, Gotlib IH. Neural correlates of rumination in depression. Cogn Affect Behav Neurosci. 2010 Dec;10(4):470–8.

37 Liston C, Chen AC, Zebley BD, Drysdale AT, Gordon R, Leuchter B, et al. Default mode network mechanisms of transcranial magnetic stimulation in depression. Biol Psychiatry. 2014 Oct;76(7):517–26.

38 Goya-Maldonado R, Brodmann K, Keil M, Trost S, Dechent P, Gruber O. Differentiating unipolar and bipolar depression by alterations in large-scale brain networks. Hum Brain Mapp. 2016 Feb;37(2):808–18.

39 Weigand A, Horn A, Caballero R, Cooke D, Stern AP, Taylor SF, et al. Prospective Validation That Subgenual Connectivity Predicts Antidepressant Efficacy of Transcranial Magnetic Stimulation Sites. Biol Psychiatry. 2018 Jul;84(1):28–37.

40 Fox MD, Liu H, Pascual-Leone A. Identification of reproducible individualized targets for treatment of depression with TMS based on intrinsic connectivity. Neuroimage. 2013 Feb;66:151–60.

41 Fox MD, Buckner RL, White MP, Greicius MD, Pascual-Leone A. Efficacy of transcranial magnetic stimulation targets for depression is related to intrinsic functional connectivity with the subgenual cingulate. Biol Psychiatry. 2012 Oct;72(7):595–603.

42 Mayberg HS, Lozano AM, Voon V, McNeely HE, Seminowicz D, Hamani C, et al. Deep brain stimulation for treatment-resistant depression. Neuron. 2005 Mar;45(5):651–60.

43 Drevets WC, Savitz J, Trimble M. The subgenual anterior cingulate cortex in mood disorders. CNS Spectr. 2008 Aug;13(8):663–81.

44 Baeken C, Duprat R, Wu GR, De Raedt R, van Heeringen K. Subgenual Anterior Cingulate-Medial Orbitofrontal Functional Connectivity in Medication-Resistant Major Depression: A Neurobiological Marker for Accelerated Intermittent Theta Burst Stimulation Treatment? Biol Psychiatry Cogn Neurosci Neuroimaging. 2017 Oct;2(7):556–65.

45 Ge R, Blumberger DM, Downar J, Daskalakis ZJ, Dipinto AA, Tham JC, et al. Abnormal functional connectivity within resting-state networks is related to rTMS-based therapy effects of treatment resistant depression: A pilot study. J Affect Disord. 2017 Aug;218(6):75–81.

46 Philip NS, Barredo J, van 't Wout-Frank M, Tyrka AR, Price LH, Carpenter LL. Network Mechanisms of Clinical Response to Transcranial Magnetic Stimulation in Posttraumatic Stress Disorder and Major Depressive Disorder. Biol Psychiatry. 2018 Feb;83(3):263–72.

47 Kang JI, Lee H, Jhung K, Kim KR, An SK, Yoon KJ, et al. Frontostriatal Connectivity Changes in Major Depressive Disorder After Repetitive Transcranial Magnetic Stimulation: A Randomized Sham-Controlled Study. J Clin Psychiatry. 2016 Sep;77(9):e1137–43.

48 Avissar M, Powell F, Ilieva I, Respino M, Gunning FM, Liston C, et al. Functional connectivity of the left DLPFC to striatum predicts treatment response of depression to TMS. Brain Stimul. 2017 Sep - Oct;10(5):919–25.

49 Bakker N, Shahab S, Giacobbe P, Blumberger DM, Daskalakis ZJ, Kennedy SH, et al. rTMS of the dorsomedial prefrontal cortex for major depression: safety, tolerability, effectiveness, and outcome predictors for 10 Hz versus intermittent theta-burst stimulation. Brain Stimul. 2015 Mar-Apr;8(2):208–15.

50 Peinemann A, Lehner C, Conrad B, Siebner HR. Age-related decrease in paired-pulse intracortical inhibition in the human primary motor cortex. Neurosci Lett. 2001 Nov;313(1-2):33–6.

51 Pitcher JB, Ogston KM, Miles TS. Age and sex differences in human motor cortex input-output characteristics. J Physiol. 2003 Jan;546(Pt 2):605–13.

52 Quentin R, Chanes L, Migliaccio R, Valabrègue R, Valero-Cabré A. Fronto-tectal white matter connectivity mediates facilitatory effects of non-invasive neurostimulation on visual detection. Neuroimage. 2013 Nov;82:344–54.

53 Quentin R, Chanes L, Vernet M, Valero-Cabré A. Fronto-Parietal Anatomical Connections Influence the Modulation of Conscious Visual Perception by High-Beta Frontal Oscillatory Activity. Cereb Cortex. 2015 Aug;25(8):2095–101.

54 Quentin R, Elkin Frankston S, Vernet M, Toba MN, Bartolomeo P, Chanes L, et al. Visual Contrast Sensitivity Improvement by Right Frontal High-Beta Activity Is Mediated by Contrast Gain Mechanisms and Influenced by Fronto-Parietal White Matter Microstructure. Cereb Cortex. 2016 Jun;26(6):2381–90.

55 Kleim JA, Chan S, Pringle E, Schallert K, Procaccio V, Jimenez R, et al. BDNF val66met polymorphism is associated with modified experience-dependent plasticity in human motor cortex. Nat Neurosci. 2006 Jun;9(6):735–7.

56 Cheeran B, Talelli P, Mori F, Koch G, Suppa A, Edwards M, et al. A common polymorphism in the brain-derived neurotrophic factor gene (BDNF) modulates human cortical plasticity and the response to rTMS. J Physiol. 2008 Dec;586(23):5717–25.

57 Stokes MG, Chambers CD, Gould IC, Henderson TR, Janko NE, Allen NB, et al. Simple metric for scaling motor threshold based on scalp-cortex distance: application to studies using transcranial magnetic stimulation. J Neurophysiol. 2005 Dec;94(6):4520–7.

58 Fitzgerald PB, Maller JJ, Hoy KE, Thomson R, Daskalakis ZJ. Exploring the optimal site for the localization of dorsolateral prefrontal cortex in brain stimulation experiments. Brain Stimul. 2009 Oct;2(4):234–7.

59 Dubois J, Adolphs R. Building a Science of Individual Differences from fMRI. Trends Cogn Sci. 2016 Jun;20(6):425–43.

60 Van Horn JD, Grafton ST, Miller MB. Individual Variability in Brain Activity: A Nuisance or an Opportunity? Brain Imaging Behav. 2008 Dec;2(4):327–34.

61 Miller MB, Van Horn JD. Individual variability in brain activations associated with episodic retrieval: a role for large-scale databases. Int J Psychophysiol. 2007 Feb;63(2):205–13.

62 Silvanto J, Muggleton NG. New light through old windows: moving beyond the "virtual lesion" approach to transcranial magnetic stimulation. Neuroimage. 2008 Jan;39(2):549–52.

63 Silvanto J, Muggleton N, Walsh V. State-dependency in brain stimulation studies of perception and cognition. Trends Cogn Sci. 2008 Dec;12(12):447–54.

64 Silvanto J, Muggleton NG, Cowey A, Walsh V. Neural activation state determines behavioral susceptibility to modified theta burst transcranial magnetic stimulation. Eur J Neurosci. 2007 Jul;26(2):523–8.

65 Singh A, Erwin-Grabner T, Sutcliffe G, Antal A, Paulus W, Goya-Maldonado R. Personalized repetitive transcranial magnetic stimulation temporarily alters default mode network in healthy subjects. Sci Rep. 2019 Apr;9(1):5631.

66 Creutzfeldt OD, Fromm GH, Kapp H. Influence of transcortical d-c currents on cortical neuronal activity. Exp Neurol. 1962 Jun;5(6):436–52.

67 Bindman LJ, Lippold OC, Redfearn JW. The action of brief polarizing currents on the cerebral cortex of the rat (1) during current flow and (2) in the production of long-lasting after-effects. J Physiol. 1964 Aug;172(3):369–82.

68 Woods AJ, Antal A, Bikson M, Boggio PS, Brunoni AR, Celnik P, et al. A technical guide to tDCS, and related non-invasive brain stimulation tools. Clin Neurophysiol. 2016 Feb;127(2):1031–48.

69 Nitsche MA, Paulus W. Excitability changes induced in the human motor cortex by weak transcranial direct current stimulation. J Physiol. 2000 Sep;527(Pt 3):633–9.

70 Nitsche MA, Paulus W. Sustained excitability elevations induced by transcranial DC motor cortex stimulation in humans. Neurology. 2001 Nov;57(10):1899–901.

71 Nitsche MA, Paulus W. Transcranial direct current stimulation—update 2011. Restor Neurol Neurosci. 2011;29(6):463–92.

72 Been G, Ngo TT, Miller SM, Fitzgerald PB, editors. The use of tDCS and CVS as methods of non-invasive brain stimulation. In: Brain Research Reviews, Vol. 56. Elsevier; 2007. pp. 346–61.

73 Doruk D, Gray Z, Bravo GL, Pascual-Leone A, Fregni F. Effects of tDCS on executive function in Parkinson's disease. Neurosci Lett. 2014 Oct;582:27–31.

74 Flöel A. tDCS-enhanced motor and cognitive function in neurological diseases. Neuroimage. 2014 Jan;85(Pt 3):934–47.

75 Hsu WY, Ku Y, Zanto TP, Gazzaley A. Effects of noninvasive brain stimulation on cognitive function in healthy aging and Alzheimer's disease: a systematic review and meta-analysis. Neurobiol Aging. 2015 Aug;36(8):2348–59.

76 Jacobson L, Koslowsky M, Lavidor M. tDCS polarity effects in motor and cognitive domains: a meta-analytical review. Exp Brain Res. 2012 Jan;216(1):1–10.

77 Tremblay S, Lepage JF, Latulipe-Loiselle A, Fregni F, Pascual-Leone A, Théoret H. The uncertain outcome of prefrontal tDCS. Brain Stimul. 2014 Nov-Dec;7(6):773–83.

78 Prehn K, Flöel A. Potentials and limits to enhance cognitive functions in healthy and pathological aging by tDCS. Front Cell Neurosci. 2015 Sep;9:355.

79 Roe JM, Nesheim M, Mathiesen NC, Moberget T, Alnæs D, Sneve MH. The effects of tDCS upon sustained visual attention are dependent on cognitive load. Neuropsychologia. 2016 Jan;80:1–8.

80 Horvath JC, Vogrin SJ, Carter O, Cook MJ, Forte JD. Effects of a common transcranial direct current stimulation (tDCS) protocol on motor evoked potentials found to be highly variable within individuals over 9 testing sessions. Exp Brain Res. 2016 Sep;234(9):2629–42.

81 Jung-Hoon Kim. Do-Won Kim, Won-Hyuk Chang, Yun-Hee Kim, Chang-Hwan Im. Inconsistent outcomes of transcranial direct current stimulation (tDCS) may be originated from the anatomical differences among individuals: a simulation study using individual MRI data. In: 2013 35th Annual International Conference of the IEEE Engineering in Medicine and Biology Society (EMBC). IEEE; 2013. p. 823–5.

82 Laakso I, Tanaka S, Koyama S, De Santis V, Hirata A. Inter-subject Variability in Electric Fields of Motor Cortical tDCS. Brain Stimul. 2015 Sep-Oct;8(5):906–13.

83 Madhavan S, Sriraman A, Freels S, Madhavan S, Sriraman A, Freels S. Reliability and Variability of tDCS Induced Changes in the Lower Limb Motor Cortex. Brain Sci. 2016 Jul;6(3):26.

84 Buch ER, Santarnecchi E, Antal A, Born J, Celnik PA, Classen J, et al. Effects of tDCS on motor learning and memory formation: A consensus and critical position paper. Clin Neurophysiol. 2017 Apr;128(4):589–603.

85 Wörsching J, Padberg F, Helbich K, Hasan A, Koch L, Goerigk S, et al. Test-retest reliability of prefrontal transcranial Direct Current Stimulation (tDCS) effects on functional MRI connectivity in healthy subjects. Neuroimage. 2017 Jul;155:187–201.

86 Bikson M, Rahman A, Datta A. Computational models of transcranial direct current stimulation. Clin EEG Neurosci. 2012 Jul;43(3):176–83.

87 Abdel Rahman AM, Ryczko M, Pawling J, Dennis JW. Probing the hexosamine biosynthetic pathway in human tumor cells by multitargeted tandem mass spectrometry. ACS Chem Biol. 2013 Sep;8(9):2053–62.

88 Bikson M, Name A, Rahman A. Origins of specificity during tDCS: anatomical, activity-selective, and input-bias mechanisms. Front Hum Neurosci. 2013 Oct;7:688.

89 Rahman A, Reato D, Arlotti M, Gasca F, Datta A, Parra LC, et al. Cellular effects of acute direct current stimulation: somatic and synaptic terminal effects. J Physiol. 2013 May;591(10):2563–78.

90 Rahman A, Toshev PK, Bikson M. Polarizing cerebellar neurons with transcranial Direct Current Stimulation. Clin Neurophysiol. 2014 Mar;125(3):435–8.

91 Datta A, Bansal V, Diaz J, Patel J, Reato D, Bikson M. Gyri-precise head model of transcranial direct current stimulation: improved spatial focality using a ring electrode versus conventional rectangular pad. Brain Stimul. 2009 Oct;2(4):201–7.

92 Radman T, Datta A, Ramos RL, Brumberg JC, Bikson M. One-dimensional representation of a neuron in a uniform electric field. In: 2009 Annual International Conference of the IEEE Engineering in Medicine and Biology Society. IEEE; 2009. p. 6481–4.

93 Nitsche MA, Nitsche MS, Klein CC, Tergau F, Rothwell JC, Paulus W. Level of action of cathodal DC polarisation induced inhibition of the human motor cortex. Clin Neurophysiol. 2003 Apr;114(4):600–4.

94 Batsikadze G, Moliadze V, Paulus W, Kuo MF, Nitsche MA. Partially non-linear stimulation intensity-dependent effects of direct current stimulation on motor cortex excitability in humans. J Physiol. 2013 Apr;591(7):1987–2000.

95 Monte-Silva K, Kuo MF, Hessenthaler S, Fresnoza S, Liebetanz D, Paulus W, et al. Induction of late LTP-like plasticity in the human motor cortex by repeated non-invasive brain stimulation. Brain Stimul. 2013 May;6(3):424–32.

96 Antal A, Terney D, Poreisz C, Paulus W. Towards unravelling task-related modulations of neuroplastic changes induced in the human motor cortex. Eur J Neurosci. 2007 Nov;26(9):2687–91.

97 Paulus W, Rothwell JC. Membrane resistance and shunting inhibition: where biophysics meets state-dependent human neurophysiology. J Physiol. 2016 May;594(10):2719–28.

98 Bikson M, Datta A, Rahman A, Scaturro J. Electrode montages for tDCS and weak transcranial electrical stimulation: role of "return" electrode's position and size. Clin Neurophysiol. 2010 Dec;121(12):1976–8.

99 Bikson M, Datta A. Guidelines for precise and accurate computational models of tDCS. Brain Stimul. 2012 Jul;5(3):430–1.

100 Fischer DB, Fried PJ, Ruffini G, Ripolles O, Salvador R, Banus J, et al. Multifocal tDCS targeting the resting state motor network increases cortical excitability beyond traditional tDCS targeting unilateral motor cortex. Neuroimage. 2017 Aug;157:34–44.

101 Villamar MF, Volz MS, Bikson M, Datta A, Dasilva AF, Fregni F. Technique and considerations in the use of 4x1 ring high-definition transcranial direct current stimulation (HD-tDCS). J Vis Exp. 2013 Jul;(77):e50309.

102 Alam M, Truong DQ, Khadka N, Bikson M. Spatial and polarity precision of concentric high-definition transcranial direct current stimulation (HD-tDCS). Phys Med Biol. 2016 Jun;61(12):4506–21.

103 Bai S, Dokos S, Ho KA, Loo C. A computational modelling study of transcranial direct current stimulation montages used in depression. Neuroimage. 2014 Feb;87:332–44.

104 Palm U, Hasan A, Strube W, Padberg F. tDCS for the treatment of depression: a comprehensive review. Eur Arch Psychiatry Clin Neurosci. 2016 Dec;266(8):681–94.

105 Brunoni AR, Moffa AH, Sampaio-Junior B, Borrione L, Moreno ML, Fernandes RA, et al.; ELECT-TDCS Investigators. Trial of Electrical Direct-Current Therapy versus Escitalopram for Depression. N Engl J Med. 2017 Jun;376(26):2523–33.

106 Brunoni AR, Padberg F, Vieira EL, Teixeira AL, Carvalho AF, Lotufo PA, et al. Plasma biomarkers in a placebo-controlled trial comparing tDCS and escitalopram efficacy in major depression. Prog Neuropsychopharmacol Biol Psychiatry. 2018 Aug;86:211–7.

107 Sampaio-Junior B, Tortella G, Borrione L, Moffa AH, Machado-Vieira R, Cretaz E, et al. Efficacy and Safety of Transcranial Direct Current Stimulation as an Add-on Treatment for Bipolar Depression: A Randomized Clinical Trial. JAMA Psychiatry. 2018 Feb;75(2):158–66.

108 Welch ES, Weigand A, Hooker JE, Philip NS, Tyrka AR, Press DZ, et al. Feasibility of Computerized Cognitive-Behavioral Therapy Combined With Bifrontal Transcranial Direct Current Stimulation for Treatment of Major Depression. Neuromodulation. 2018. DOI: 10.1111/ner.12807.

109 Keeser D, Meindl T, Bor J, Palm U, Pogarell O, Mulert C, et al. Prefrontal transcranial direct current stimulation changes connectivity of resting-state networks during fMRI. J Neurosci. 2011 Oct;31(43):15284–93.

110 Keeser D, Padberg F, Reisinger E, Pogarell O, Kirsch V, Palm U, et al. Prefrontal direct current stimulation modulates resting EEG and event-related potentials in healthy subjects: a standardized low resolution tomography (sLORETA) study. Neuroimage. 2011 Mar;55(2):644–57.

111 Valiengo L, Benseñor IM, Goulart AC, de Oliveira JF, Zanao TA, Boggio PS, et al. The sertraline versus electrical current therapy for treating depression clinical study (select-TDCS): results of the crossover and follow-up phases. Depress Anxiety. 2013 Jul;30(7):646–53.

112 Brunoni AR, Machado-Vieira R, Sampaio-Junior B, Vieira EL, Valiengo L, Benseñor IM, et al. Plasma levels of soluble TNF receptors 1 and 2 after tDCS and sertraline treatment in major depression: results from the SELECT-TDCS trial. J Affect Disord. 2015 Oct;185:209–13.

113 Brunoni AR, Tortella G, Benseñor IM, Lotufo PA, Carvalho AF, Fregni F. Cognitive effects of transcranial direct current stimulation in depression: results from the SELECT-TDCS trial and insights for further clinical trials. J Affect Disord. 2016 Sep;202:46–52.

114 Nitsche MA, Kuo MF, Karrasch R, Wächter B, Liebetanz D, Paulus W. Serotonin affects transcranial direct current-induced neuroplasticity in humans. Biol Psychiatry. 2009 Sep;66(5):503–8.

115 Kuo HI, Paulus W, Batsikadze G, Jamil A, Kuo MF, Nitsche MA. Chronic Enhancement of Serotonin Facilitates Excitatory Transcranial Direct Current Stimulation-Induced Neuroplasticity. Neuropsychopharmacology. 2016 Apr;41(5):1223–30.

116 Dell'Osso B, Dobrea C, Arici C, Benatti B, Ferrucci R, Vergari M, et al. Augmentative transcranial direct current stimulation (tDCS) in poor responder depressed patients: a follow-up study. CNS Spectr. 2014 Aug;19(4):347–54.

117 Martin DM, Alonzo A, Ho KA, Player M, Mitchell PB, Sachdev P, et al. Continuation transcranial direct current stimulation for the prevention of relapse in major depression. J Affect Disord. 2013 Jan;144(3):274–8.

118 McClintock SM, Reti IM, Carpenter LL, McDonald WM, Dubin M, Taylor SF, et al.; National Network of Depression Centers rTMS Task Group; American Psychiatric Association Council on Research Task Force on Novel Biomarkers and Treatments. Consensus Recommendations for the Clinical Application of Repetitive Transcranial Magnetic Stimulation (rTMS) in the Treatment of Depression. J Clin Psychiatry. 2018 Jan/Feb;79(1):35–48.

119 Siddiqi SH, Trapp NT, Shahim P, Hacker CD, Laumann TO, Kandala S, et al. Individualized Connectome-Targeted Transcranial Magnetic Stimulation for Neuropsychiatric Sequelae of Repetitive Traumatic Brain Injury in a Retired NFL Player. J Neuropsychiatry Clin Neurosci. 2019. DOI: 10.1176/appi.neuropsych.18100230.

120 Siddiqi SH, Trapp NT, Hacker CD, Laumann TO, Kandala S, Hong X, et al. Repetitive Transcranial Magnetic Stimulation with Resting-State Network Targeting for Treatment-Resistant Depression in Traumatic Brain Injury: A Randomized, Controlled, Double-Blinded Pilot Study. J Neurotrauma. 2019 Apr;36(8):1361–74.

121 Luber BM, Davis S, Bernhardt E, Neacsiu A, Kwapil L, Lisanby SH, et al. Using neuroimaging to individualize TMS treatment for depression: toward a new paradigm for imaging-guided intervention. Neuroimage. 2017 Mar;148:1–7.

122 Neacsiu AD, Luber BM, Davis SW, Bernhardt E, Strauman TJ, Lisanby SH. On the Concurrent Use of Self-System Therapy and Functional Magnetic Resonance Imaging-Guided Transcranial Magnetic Stimulation as Treatment for Depression. J ECT. 2018 Dec;34(4):266–73.

123 Sathappan AV, Luber BM, Lisanby SH. The Dynamic Duo: combining noninvasive brain stimulation with cognitive interventions. Prog Neuropsychopharmacol Biol Psychiatry. 2019 Mar;89:347–60.

124 Nord CL, Halahakoon DC, Limbachya T, Charpentier C, Lally N, Walsh V, et al. Neural predictors of treatment response to brain stimulation and psychological therapy in depression: a double-blind randomized controlled trial. Neuropsychopharmacology. 2019 Aug;44(9):1613–22.

125 Filmer HL, Ehrhardt SE, Bollmann S, Mattingley JB, Dux PE. Accounting for individual differences in the response to tDCS with baseline levels of neurochemical excitability. Cortex. 2019 Jun;115:324–34.

126 Bajbouj M, Aust S, Spies J, Herrera-Melendez AL, Mayer SV, Peters M, et al. PsychotherapyPlus: augmentation of cognitive behavioral therapy (CBT) with prefrontal transcranial direct current stimulation (tDCS) in major depressive disorder-study design and methodology of a multicenter double-blind randomized placebo-controlled trial. Eur Arch Psychiatry Clin Neurosci. 2018 Dec;268(8):797–808.

127 Brunoni AR, Valiengo L, Baccaro A, Zanão TA, de Oliveira JF, Goulart A, et al. The sertraline vs. electrical current therapy for treating depression clinical study: results from a factorial, randomized, controlled trial. JAMA Psychiatry. 2013 Apr;70(4):383–91.

4 Personalized repetitive transcranial magnetic stimulation temporarily alters default mode network in healthy subjects

Despite increasing clinical use, the basic neural mechanism of action of 10 Hz rTMS protocol remains little explored in healthy subjects. As we discussed in chapter 3, a lot of variability in rTMS response exists, likely because of inaccurate brain targeting. This variability can therefore be tackled by more accurate brain targeting based on individual subject's rsfMRI. Considering, healthy human brains are devoid of the confounding effects of depressive episodes, they can better inform researchers about further rTMS optimizations that may be more easily translated into depressed brains for more targeted treatment protocols. A handful of studies have investigated the effects of rTMS in healthy volunteers, but the mechanism of action of a single complete session (3000 pulses) of 10 Hz rTMS remains elusive.

The DMN has been implicated in the pathophysiology of depression and its hyperconnectivity to the sgACC has been reported to be "normalized" in patients after successful rTMS treatment [146, 250]. Previous studies have retrospectively revealed that the antidepressant response of left DLPFC stimulation is correlated to the negative functional connectivity of the stimulation location and the sgACC [163, 165], further reiterating the use of neuroimaging techniques to select stimulation sites [239]. Here, we aimed to examine the mechanism of action of a full single session of 10 Hz rTMS. We developed and prospectively validated a method to select individual left DLPFC stimulation sites as targets and precisely deliver stimulation at these targets with assistance of online neuronavigation. For this, we computed the spatially independent components using FSL and identified those that best spanned the left DLPFC and the sgACC regions to identify the strongest node at left DLPFC that is in anticorrelation to the DMN. By accommodating the inter-individual variation in functional connectome by means of rsfMRI, we personalize HF-rTMS stimulation sites in healthy volunteers. Upon stimulation at the personalized left DLPFC site in healthy subjects, we evaluated the pre- and post-stimulation effects on the DMN through multiple rsfMRI windows using a double blind (interviewer and participant), crossover, and sham controlled study design.

The insights from one session of 10 Hz rTMS stimulation in healthy subjects will be relevant for a better understanding of its acute effects, eliciting progress towards more robust rTMS interventions. Our study applying a full single dose of 10 Hz rTMS in healthy subjects with a personalized approach and the approximately one hour follow up using rsfMRI post stimulation are the novel features of this work. This longer observation period could conceivably reveal pertinent insights into the temporal dynamics of action of 10 Hz rTMS. If the basic mechanism of DMN after a single dose of 10 Hz rTMS at left DLPFC in healthy volunteers is analogous to that reported after multiple sessions of 10 Hz rTMS in patient population, we hypothesize that a full 10 Hz rTMS session would change the functional connectivity of the sgACC to the DMN. As healthy subjects have normal baseline connectivity, we expect a decrease in functional connectivity of the sgACC to DMN. As reviewed in Chapter 1, several personality dimensions have been reported to predict antidepressant effects and rTMS response. The HA dimension of the TCI has been related to the activity of sgACC in adolescents [251], healthy adults [252, 253] and to pharmacological antidepressant responses in adults [184, 254, 255]. Therefore, we selected this dimension of personality to explore if HA would be related to modulatory effects from 10 Hz rTMS and hypothesized that HA scores would correlate with DMN changes at the sgACC. In addition, to explore the potential behavioural effects prompted by a single session of 10 Hz rTMS, we used Positive and Negative Affect Schedule (PANAS) to capture the mood of the subjects prior and subsequent to the stimulation. As previous studies have shown associations between activity in sgACC and sad mood [256, 257] as well as rumination [148] in healthy subjects, we speculated that any decoupling of sgACC and DMN resulting from rTMS would cause a decrease in the self-rated negative affect of participants. Further details on the used methodology, the outcomes of the experiments and a corresponding discussion of the results are presented in the next section.

Individual author contributions are mentioned at the end of the chapter. The article was published under Creative Commons Attribution 4.0 International License.

SCIENTIFIC REP⚙RTS

OPEN

Personalized repetitive transcranial magnetic stimulation temporarily alters default mode network in healthy subjects

Received: 4 December 2018
Accepted: 19 March 2019
Published online: 04 April 2019

Aditya Singh[1], Tracy Erwin-Grabner[1], Grant Sutcliffe[1], Andrea Antal[2,3], Walter Paulus[2] & Roberto Goya-Maldonado[1]

High frequency repetitive transcranial magnetic stimulation (HF-rTMS) delivered to the left dorsolateral prefrontal cortex (DLPFC) is an effective treatment option for treatment resistant depression. However, the underlying mechanisms of a full session of HF-rTMS in healthy volunteers have not yet been described. Here we investigated, with a personalized selection of DLPFC stimulation sites, the effects driven by HF-rTMS in healthy volunteers (n = 23) over the default mode network (DMN) in multiple time windows. After a complete 10 Hz rTMS (3000 pulses) session, we observe a decrease of functional connectivity between the DMN and the subgenual Anterior Cingulate Cortex (sgACC), as well as the ventral striatum (vStr). A negative correlation between the magnitude of this decrease in the right sgACC and the harm avoidance domain measure from the Temperament and Character Inventory was observed. Moreover, we identify that coupling strength of right vStr with the DMN post-stimulation was proportional to a decrease in self-reports of negative mood from the Positive and Negative Affect Schedule. This shows HF-rTMS attenuates perception of negative mood in healthy recipients in agreement with the expected effects in patients. Our study, by using a personalized selection of DLPFC stimulation sites, contributes understanding the effects of a full session of rTMS approved for clinical use in depression over related brain regions in healthy volunteers.

Despite its increasing relevance for clinical use, the basic underlying mechanism of action of repetitive transcranial magnetic stimulation (rTMS) protocols remains unexplored in healthy subjects for a complete session of 10 Hz rTMS (3000 pulses in 37.5 min), which is the first FDA approved treatment protocol for treatment resistant depression. Much variability in the resulting rTMS response exists, which could be reduced by accurate brain targeting based upon individual subject's resting state functional magnetic resonance imaging (rsfMRI). Independent from the confounding effects of depression related symptomatology, an increased comprehension of the underlying neural mechanisms of a full session of HF-rTMS in a healthy human brain will help to better inform future research into the depressed brain and its treatment. While several studies have investigated the effects of rTMS in healthy volunteers[1–4], the mechanism of action of the complete session of 10 Hz rTMS with up to 1 hour observation of neural effects has not yet been performed.

The default mode network (DMN) is a network consisting of the medial and lateral temporal lobe, medial prefrontal cortex and medial and lateral parietal cortex. It has been implicated in the pathophysiology of depression[5–17] and its hyperconnectivity to the subgenual Anterior Cingulate Cortex (sgACC) has been reported to be normalized after successful rTMS treatment[5,18]. A recent study[19] explored in healthy subjects the mechanism of about a third of the 10 Hz rTMS protocol (1200 pulses in 10 min) typically used to treat depression. The authors reported an increase in the functional connectivity of the sgACC to another network, consisting of the dorsal cingulate cortex, posterior dorsomedial prefrontal cortex, dorsolateral prefrontal cortex (DLPFC), inferior parietal lobule, inferior frontal cortex and posterior temporal lobes, but no changes to a network resembling the DMN.

[1]Systems Neuroscience and Imaging in Psychiatry, Department of Psychiatry and Psychotherapy of the University Medical Center Göttingen, Göttingen, Germany. [2]Department of Clinical Neurophysiology of the University Medical

Here, we aimed to address the mechanism of action of a complete session of 10 Hz rTMS, targeting individually selected left DLPFC stimulation sites and precisely assisted by online neuronavigation.

Fitzgerald and colleagues[20] investigated different methods of targeting and stimulating the DLPFC. They concluded that targets defined on the 10–20 electroencephalographic (EEG) system or those selected with the use of individual structural MRI images and a neuronavigation system are spatially more precise than those selected with the "standard procedure", referring to the scalp location 5 cm anterior to the motor cortex[21]. Additionally reinforcing the use of neuroimaging techniques to select stimulation sites, studies revealed that the antidepressant response of left DLPFC stimulation can be predicted by the negative functional connectivity with the sgACC[22,23]. Therefore, to personalize HF-rTMS stimulation sites in healthy volunteers, we integrated the individual variation captured by rsfMRI[24] in a protocol that combined spatial and temporal information from independent components that best represented the left DLPFC and the sgACC regions.

In this study we evaluated the DMN by using multiple rsfMRI windows before and after stimulation in a double blind (interviewer and participant), crossover, and sham controlled study with healthy subjects. We argue that insights from one session of 10 Hz rTMS stimulation in healthy subjects can be relevant for a better understanding of the HF-rTMS antidepressant treatment, further prompting development of more efficient rTMS interventions. Such an application of single dosage of HF-rTMS in healthy subjects is unprecedented and the ~1 hour follow up using rsfMRI post stimulation will expectedly reveal pertinent insights in to the action of HF-rTMS. Reasoning that healthy subjects have normal baseline connectivity, we hypothesized that a full HF-rTMS session would change the functional connectivity of the sgACC to the DMN. If the underlying mechanism of 10 Hz rTMS at left DLPFC in healthy volunteers is analogous to that reported in patient population, then we would expect to see a decrease in functional connectivity of the sgACC to DMN. The harm avoidance (HA) dimension of the Temperament and Character Inventory (TCI) has been related to the activity of sgACC in adolescents[25], healthy adults[26,27] and to pharmacological antidepressant responses in adults[28–30]. Therefore, we selected this dimension of personality, as well as the Positive and Negative Affect Schedule (PANAS) to investigate how HF-rTMS driven effects would relate to phenotypic information from the healthy participants. We hypothesized that HA scores would anticipate DMN changes reaching the sgACC. Lastly, in agreement with previous studies that have shown that activity in sgACC is associated with sad mood[25,31,32] and rumination[13] in healthy subjects, we speculated that rTMS induced decoupling of sgACC and DMN would result in a decrease in the self-rated negative affect of participants after stimulation.

Results

Only 1 subject was excluded due to not tolerating the stimulation, therefore data from 23 subjects were included in the final analysis. The mean age of the subjects was 25.8 ± 5.5 years with 9 female subjects.

Target reproducibility. To establish the consistency of the rsfMRI based method for personalized target selection, one must first test whether target sites generated from different datasets of the same subject will concur. To test the reproducibility of our target selection process, we repeated the day 1 steps of target selection also on pre-stimulation rsfMRI data from the day on which real stimulation was delivered. We then calculated the Euclidean distances between targets selected from baseline and stimulation day. The mean distance between the two was 10.9 mm (6.74 mm SD), which was significantly less than 20 mm (two tailed, one sample t test, t value $= -6.61$, $p < 0.0001$) as presupposed in the literature (see Discussion). Figure 1, upper panel presents a bar plot of the distances between the targets for each subject. There are only three subjects for whom the distances between the day 1 and day 2 targets are slightly above 20 mm (diameter of the electric field sphere covered by figure-of-eight coils)[33]. Figure 1, lower panel, presents a qualitative comparison of the modelled electric field (SimNIBS[34]) generated by stimulation at targets for two example subjects.

Target quality. The other important aspect of our target selection process is the quality of the target that it yields. Effective HF-rTMS stimulation for the treatment of depression using left DPLFC targets is achieved by stimulating at targets with higher negative functional connectivity to sgACC[22,23]. Thus, we calculated the functional connectivity between the targets selected by our method (named individual DLPFC, indDLPFC) and the right sgACC, and also between fixed MNI coordinates based left DLPFC (named fxdDLPFC) and the right sgACC. We used 2 mm radius ROI spheres to extract the betas from DLPFC regions. In Fig. 2, upper panel, note that indDLPFC targets spread across the left DLPFC and having a larger radius would have caused overlaps with the fxdDLPFC target, potentially diluting specific correlations with right sgACC. We preserved with 5 mm radius of the latter as in Tik *et al.*[19] for comparison. We found a higher negative connectivity between indDLPFC and right sgACC compared to that between fxdDLPFC and right sgACC (Fig. 2, lower panel), thus implying that indDLPFC can be more promising therapeutically as compared to fxdDLPFC.

Behavioural scales results. As would be expected, we did not see any differences in healthy volunteers in the MADRS, HAM-D, YMRS, PANSS, and BDI II scores between day 2 and day 3 of stimulation. However, subjects had different physical sensations arising from the two sessions (two sample t test, $t = 4.89$, $p < 0.0001$), although they expected both stimulation sessions to be equally effective (two sample t test, $t = 1.17$, $p > 0.05$, see Fig. 3), as measured by visual analogue scales. This implies that our method of blinding successfully maintained similar levels of internal expectation between real and sham rTMS even though they experienced a difference in scalp discomfort. We believe the successful blinding was a result of nullification of external expectations of subjects. Since the subjects would have intuitively expected higher scalp sensation to lead to higher rTMS effects, we instructed the subjects that our study was testing out two different stimulation coils.

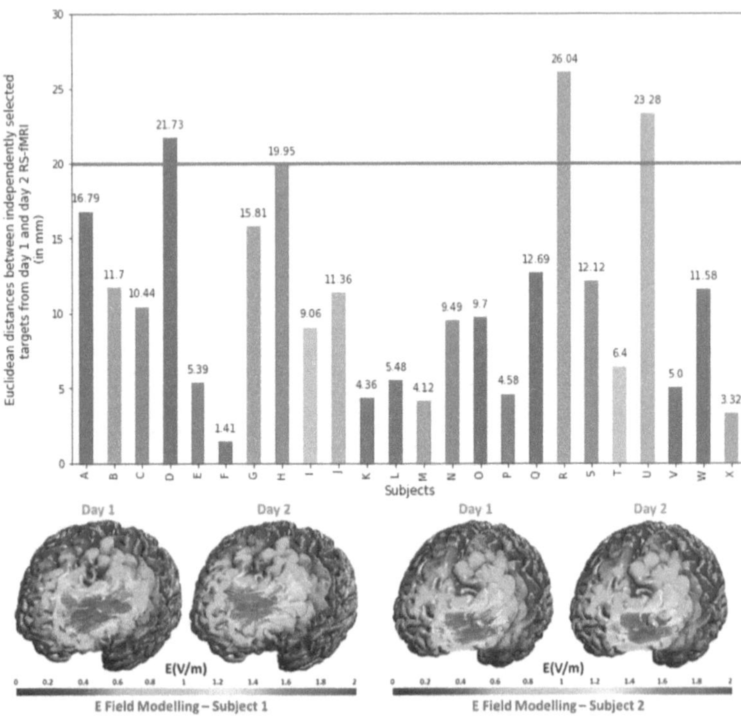

Figure 1. Distances between targets from day 1 and day 2 rsfMRI (n = 24) and electric field modelling for two example subjects. Upper panel: Bar plot of Euclidean distance between targets from day 1 rsfMRI and targets from day 2 rsfMRI. 3 subjects have a distance larger than 20 mm. Lower panel: A qualitative description of electric field modelled (Thielscher *et al.*, 2015) on the two targets from day 1 and day 2 rsfMRI data. A quick look reveals the similarity of the electric field when either of the targets are used.

wise head displacement parameters[35] from individual frames. We found that the extent of motion did not differ across rsfMRI time windows for real and sham conditions. To investigate the mechanisms of HF-rTMS over the DMN, we contrasted this network during the four rsfMRI scanning sessions in real versus sham conditions. We observed a robust decrease in functional connectivity specifically involving the sgACC (Fig. 4 a1–a3) and the ventral striatum (vStr) (Fig. 4 a4) bilaterally during the R2-R1 contrast. This decrease in functional connectivity weakened over time and the connectivity in the ventral striatum returned to baseline level already in R3, while the decrease in the sgACC coupling persisted until R3, albeit less pronounced than during R2. Also, important to note is that this decrease in functional connectivity is only seen in the real HF-rTMS stimulation condition, but not during the sham HF-rTMS.

We also explored the functional connectivity of the personalized left DLPFC stimulation site with that of sgACC as well as the DMN. The parameter estimates of the connectivity strength of personalized left DLPFC to the IC-DLPFC network and of sgACC to DMN is plotted in Fig. 5 (left y-axis). It shows that the functional connectivity of left DLPFC decreases from R0 to R1. This is followed by a decrease in the functional connectivity of the sgACC from R1 to R2. The difference in time windows during which the functional connectivities of left DLPFC and sgACC decrease, shows how the HF-rTMS effect spreads from left DLPFC to sgACC across time.

In addition, the correlation between the left DLPFC and sgACC was evaluated. The green dashed line plots the correlation coefficients (right y-axis) between the parameter estimates of left DLPFC and the sgACC during the four rsfMRI sessions (R0 to R3). It shows that the baseline negative correlation between left DLPFC and sgACC reduces during R1 but returns to more negative values during R2 and R3.

A predictor for rTMS response   The personality trait of UA has been suggested to predict therapeutic responses

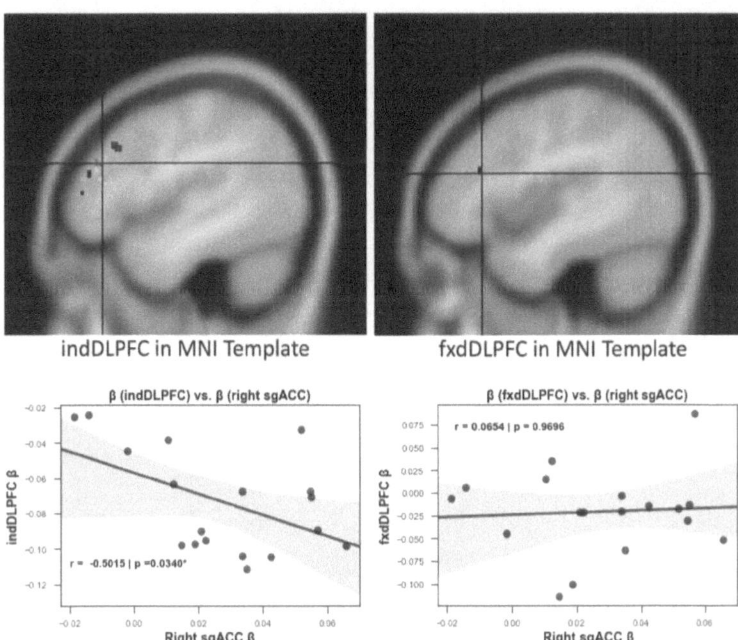

Figure 2. Left DLPFC targets and their correlation with right sgACC. Upper panel: Figure showing the 2 mm radius ROIs for targets obtained by the selection process described in the study (n = 22) [left] and the ROI of 2 mm radius around the standard MNI DLPFC coordinates [right]. Lower panel: The plot on the left shows that targets identified using the described selection process have a much higher negative correlation with right sgACC (r = −0.5015, *p < 0.05) than when the standard MNI left DLPFC coordinates are used (right plot). Since, antidepressant response of rTMS is linked to the connectivity of stimulated site and the sgACC; individualized targets would be promising for better therapeutic response than standard targets (n = 18).

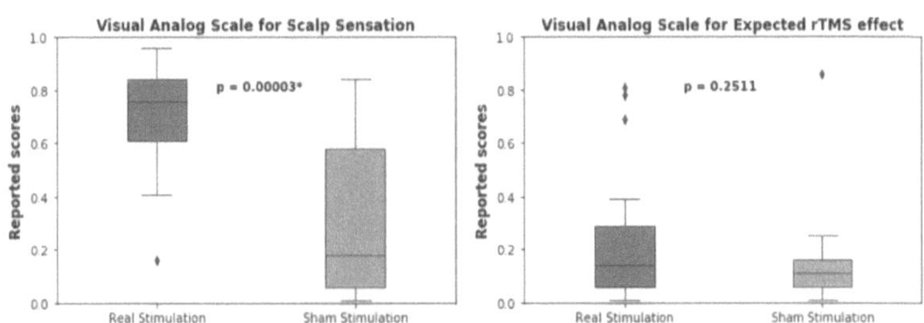

Figure 3. Physical sensation and resultant expectation about the efficacy of rTMS stimulation in subject (n = 17). The plot on the left shows that the scalp discomfort experienced by the subjects was significantly higher during the real than sham condition. However, their internal expectation of the induced changes in their mental states did not differ across real and sham stimulation sessions.

volunteers. For that, we correlated HA scores from the TCI with the parameter estimates (beta weights) changes of functional connectivity at the sgACC (R2 − R1, Fig. 6). We identified a stronger negative correlation between HA and

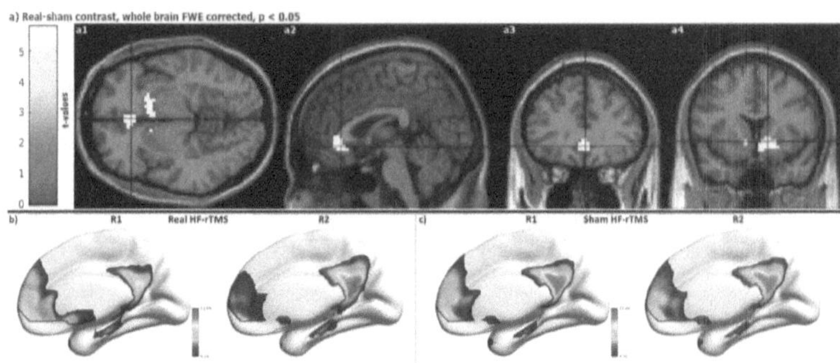

Figure 4. Functional connectivity decrease is only seen after real HF-rTMS (n = 23). (**a**) Decreased functional connectivity in bilateral sgACC (a1 to a3) and vStr (a4) during R2 when compared to R1. (**b**) Default mode network after real HF-rTMS during R1 and R2. (**c**) Default mode network after sham HF-rTMS during R1 and R2. Colorbars represent t-values.

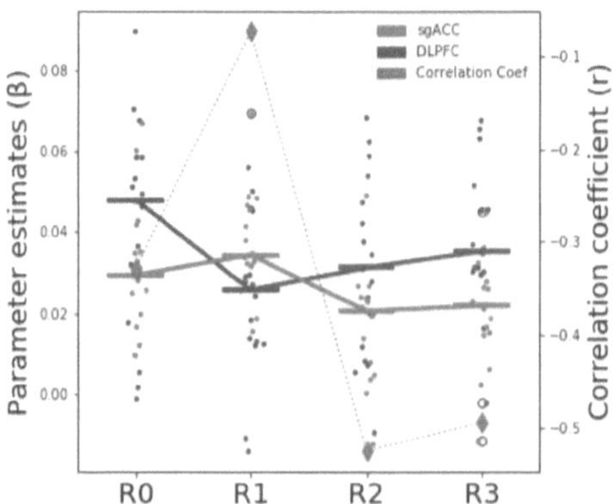

Figure 5. Plotting the interaction of personalized left DLPFC with sgACC and DMN. The plot shows the parameter estimates of left DLPFC functional connectivity to IC-DLPFC network and that of sgACC to DMN. The bars represent the median value of the parameter estimates. The plot shows how left DLPFC changes in functional connectivity occur during R1 rsfMRI session. While the sgACC functional connectivity changes follow later during R2 rsfMRI session. The green dashed line plots the correlation coefficients (right y-axis) between the parameter estimates of left DLPFC and sgACC. It shows that the baseline negative correlation between left DLPFC and sgACC is reduced after stimulation in R1 rsfMRI session, followed by stronger negative correlation between the two regions during R2 and R3 rsfMRI sessions. Refer to legend for info on color coding.

Behavioural correlates of functional connectivity changes. Considering the role of the sgACC activity in sad mood[25,31,32], we used the negative affect score from Positive and Negative Affect Schedule (PANAS), a psychometric scale that measures both positive and negative affect, to investigate behavioural outputs from real and sham rTMS. We explored the correlation between the changes in the negative affect scores (post stimulation – pre stimulation scores) with the R2 functional connectivity of the DMN to the right sgACC or to the right nucleus accumbens (NAcc). The latter was post hoc performed, since we observed robust decreases in the functional connectivity of right ventral striatum with DMN and this region is implicated in integrating emotional signals from limbic systems[37]. For that, we selected an independent coordinate for the right NAcc from the literature and extracted the beta weights. We report a trend of positive correlation (r = 0.3894, p = 0.0993) between the DMN-NAcc functional connectivity during R2 and changes in negative affect scores in the real condition, but not in the sham condition (Fig. 7).

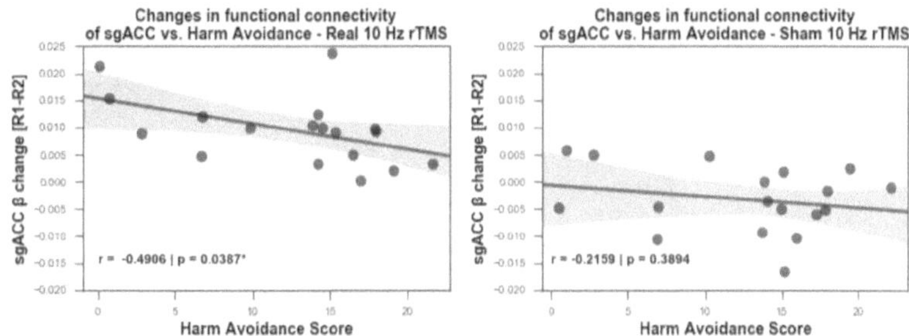

Figure 6. Harm avoidance score can predict rTMS induced reduction in functional connectivity of right sgACC (n = 18). A negative correlation between harm avoidance score and the reduction in the functional connectivity of the default mode network at the sgACC during R2 (27–32 min) from the R1 (10–15 min) rsfMRI windows, indicates that the lower the subject scores on the harm avoidance scale, the higher was the uncoupling resulting from rTMS.

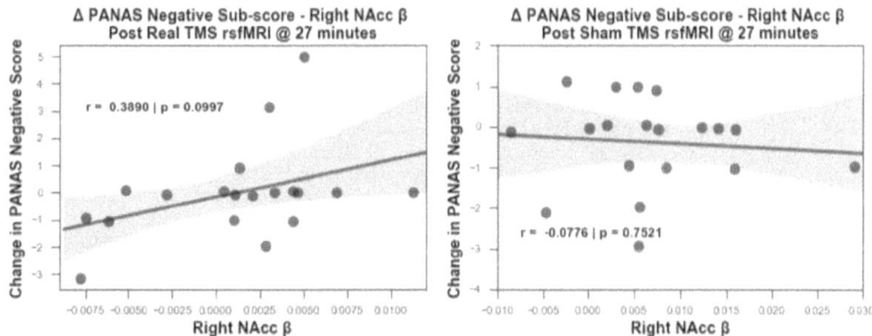

Figure 7. Subjects that perceived less negative affect after stimulation had lower functional connectivity in right NAcc (n = 19). A trend of positive correlation observed between the functional connectivity strength of the default mode network at the right NAcc and the changes in negative affect score on PANAS scale occurs in real rTMS (left) but not in sham rTMS (right) condition. The correlation shows a transient effect during R2 rsfMRI window (27–32 min) after rTMS: subjects who reported less negative emotions after stimulation showed less coupling between right NAcc and the default mode network.

Discussion

Given the lack of studies exploring the mechanism of action of single session of FDA approved 10 Hz rTMS session in healthy volunteers, it is imperative to investigate mechanisms underlying such rTMS protocols, to bridge the existing knowledge gap in our basic understanding of HF-rTMS effects. The fact that about half of the patients receiving rTMS treatment for depression still do not clinically respond[38] emphasizes the need to personalize rTMS stimulation. In an effort to improve rTMS target selection, which seems to strongly contribute to the observed variability in rTMS response[39–41], we have proposed a novel and easy to implement, rsfMRI based rTMS target selection method for the left DLPFC. As a proof of concept we have shown in our study with healthy subjects, the target selection process yields reproducible results in the same individual generated from different datasets. By means of targeting left DLPFC sites with higher negative functional connectivity to sgACC, our results show that this target selection process holds potential for increasing rTMS therapeutic response in comparison to stimulating at MNI based anatomical left DLPFC coordinates. Upon employing this target selection method to investigate the mechanism of action of a full session of 10 Hz rTMS in healthy subjects, we show a reduced functional connectivity of the DMN with the sgACC after stimulation, and report for the first time, a reduced functional connectivity of this network with the vStr (both effects peaking at 27–32 minutes post HF-rTMS, namely R2). These results are particularly important as these regions are considered core structures implicated in the pathophysiology and treatment of depression[5–7,12,13,42]. Also, our study shows that it is possible to manipulate the functional connectivity of these deeper regions by stimulating at personalized cortical targets. Moreover, we have found that the lower the harm avoidance score, the stronger the HF-rTMS effects in the sgACC 27–32 min-

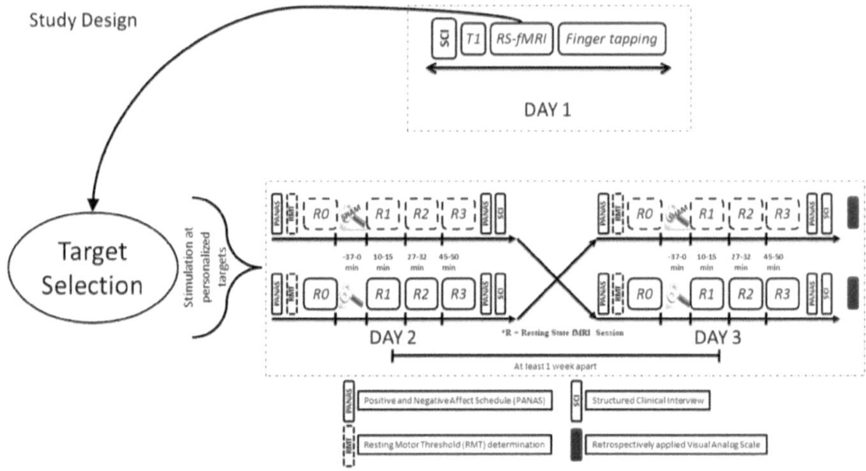

Figure 8. Study design – On day 1, after a Structured Clinical Interview (SCI) and obtaining informed consent, we acquired a T1 weighted structural image, rsfMRI and finger tapping task based fMRI. During rsfMRI, we instructed subjects to look at a fixation cross presented on a black background. The rsfMRI data was used for personalized target selection and the personalized target was then identified in T1 structural for stimulation using online neuronavigation. rsfMRI sessions were followed by an index finger tapping session (intermittent finger tapping when presented with a green or a red dot), data from which was used to determine the motor cortex location used for setting the resting motor threshold (RMT). At the beginning of experiment on day 2 and day 3, we asked each subject to complete the Positive and Negative Affect Schedule (PANAS). This was followed by resting motor threshold determination. We obtained a baseline rsfMRI scan (R0), and then delivered HF rTMS stimulation (either real or sham in a pseudo-randomized and counterbalanced way) to the subject at the pre-selected target, guided by online neuronavigation. Post HF rTMS three additional rsfMRI scans (R1, R2, and R3) were obtained over a course of 50 minutes: at 10 minutes, 27 minutes and 45 minutes to detect effects on brain resting state functional connectivity. The subjects again completed the PANAS immediately after leaving the MR scanner. This allowed us to document any short term changes in the self-rated emotional state. The subjects completed a SCI again at the end of experiment on both day 2 and day 3. To assess the effectiveness of sham blinding, we also retrospectively collected information about the perceived effects of stimulation on scalp sensation and mental state using a visual analog scale (VAS).

HF-rTMS. Lastly, we have seen that the lower the functional connectivity of the DMN to the vStr, the less negative the subjects report their emotions. This correlation pattern was not identified in the sgACC. This hints at how the influence of 10 Hz rTMS on the DMN extends to behaviour, captured by the self-report of negative affect in healthy subjects.

Previous work has highlighted that stimulation sites displaying higher negative connectivity with the sgACC can achieve stronger antidepressant effects[22,23], a characteristic that can be missed when selecting rTMS targets based on group averaging[43]. In this context and to achieve more cohesive results from stimulation in healthy volunteers, we developed a target selection process that incorporates individual functional connectivity characteristics from rsfMRI data. To establish the consistency of a method for personalized target selection in the DLPFC, one must first test whether target sites generated from different datasets of the same subject will coincide. Previous research on modelling of electric fields[33] has shown that a figure-of-eight TMS coil covers a diameter of approximately 20 mm. Therefore, our method should yield left DLPFC targets that are within 20 mm of each other. Upon calculating the Euclidean distances between targets resulting from these two independent datasets, we have shown that the left DLPFC targets from independent datasets were on average 10.9 mm (6.74 mm SD) apart from each other. This indicates that even if the target might have slightly shifted on the day of stimulation, it largely remained within the 20 mm electric field. Thus, our approach maximizes the stimulation effects by capturing the interindividual variability of networks from a previous fMRI session and guiding the stimulation coil for each volunteer to the most promising site at the left DLPFC to manipulate the sgACC. We proved this important point by showing that individualized targets selected by our method (indDLPFC) have a higher overall negative correlation than the fixed left DLPFC target (fxdDLPFC) with the right sgACC (independent coordinates)[19]. Considering the anticorrelation of the sgACC and the stimulated DLFPC site quantifies the quality of HF-rTMS response[22,23], we believe that our method can boost responses by guiding stimulation into these functionally rel-

To understand the fundamental mechanism of 10 Hz rTMS protocol we delivered a complete (3000 pulses) session of personalized 10 Hz rTMS in healthy subjects. We found a robust decrease of functional connectivity after real HF-rTMS between the DMN and the sgACC as well as the vStr during R2. This decreased coupling persists during R3 in the sgACC, albeit less pronounced, but it returns to normal in the vStr during R3. Although with a similar study design as Tik and colleagues[19], our results diverge from theirs. They reported an increase in the functional connectivity of the sgACC to a network consisting of the dorsal cingulate cortex, posterior dorsomedial prefrontal cortex, DLPFC, inferior parietal lobule, inferior frontal cortex and posterior temporal lobes. The difference in reported results potentially stems from the fact that we tracked HF-rTMS induced changes in the DMN for different and longer periods of time (during 10–15 (R1), 27–32 (R2), and 45–50 min (R3), compared to 15–21 and 31–37 min[19]) and investigated the DMN, which has been consistently described in the literature of depression[5–18,44]. We further argue that this discrepancy in results stems from the difference of networks that show an altered functional connectivity to sgACC. Tik et al.[19] have reported a network different from the DMN having an increased functional connectivity to sgACC (see above for the regions included in their network RSN #17). Our study, on the other hand, has identified changes involving the DMN and compared it across four sequential rsfMRI sessions. Additional differences in TMS stimulation included reduced length of the FDA approved HF-rTMS protocol (1200 vs. 3000 pulses), reduced stimulation intensity (80% vs. 110% of Resting motor threshold - RMT), and different stimulation delivery (anatomical standard left DLPFC target vs. functional connectivity based personalized left DLPFC target), all of which could have contributed to a difference in results. Additional analysis of parameter estimates of left DLPFC stimulation sites in IC-DLFPC and sgACC in DMN reveals that the connectivity of left DLPFC decreases initially after stimulation from R0 to R1. This is followed by a decrease in functional connectivity of the sgACC to the DMN during a later time window (from R1 to R2). Hence, we report the differences in temporal dynamics of HF-rTMS induced changes observed at the site of stimulation (left DLPFC) and at more distant locations (sgACC).

Studies have observed higher sgACC functional connectivity in depression and its role in antidepressant response[6,11,45–55]. Further studies have shown abnormal sgACC connectivity with the DMN during depression[9,13,44] and have reported decreased activity in sgACC in response to deep brain stimulation[52,56]. In this context, our results of decoupling the DMN and the sgACC after HF-rTMS in healthy subjects align well with such findings in patients with depression. However, since we have used one full session of HF-rTMS rather than 20 sessions, as is usually performed in clinical practice, this implication is limited in its interpretability. The NAcc, located at the vStr, is an integral component of the reward system and it was recently demonstrated[57] that a disrupted reward circuit is associated with the pathophysiology of depression. Another study reported that the level of functional connectivity between the NAcc and left DLPFC target predicts the antidepressant response of HF-rTMS in depression[42]. Given the importance of the sgACC and the NAcc regions in the pathophysiology of depression, the changes reported in these regions after a full session in healthy volunteers hold significance for elucidating the underlying neural network mechanisms that result in antidepressant effects. In contrast to our report of decrease in functional connectivity of sgACC in the active stimulation condition only, Taylor et al.[18] reported such a decrease across both sham and active HF-rTMS conditions. This discrepancy might stem from the fact that their results are from a patient population who were on a stable dose of medication, whereas ours are from healthy volunteers after a single session of personalised HF-rTMS.

In light of the changes of functional connectivity of sgACC reported here, and previous studies implicating sgACC and HA scores in healthy subjects[25–27] we examined the correlations between HA and effects on functional connectivity of sgACC. The negative correlation between the harm avoidance scores and the changes in functional connectivity in the real stimulation condition, but not in sham indicates that harm avoidance score can potentially anticipate the extent of the right sgACC rTMS response in healthy subjects. This is an interesting finding in light of previous studies[28–30,58], which reported that harm avoidance was a negative predictor of response to antidepressant treatment, i.e. patients with higher harm avoidance scores responded poorly to antidepressant treatment. Apart from this, Ward and colleagues[59] also reported a weak link between poor antidepressant response and higher scores on neuroticism, which itself has been shown to positively correlate with harm avoidance scores[60]. Further studies[61,62] also point towards a relation between high scores on neuroticism and low response to treatment of depression. Since our study is based on a cohort of healthy subjects, we believe harm avoidance of the TCI is a good proxy for neuroticism, given the positive correlation between the two personality measures. Hence, the correlation between HA and HF-rTMS effects in the sgACC reported here, warrants further clinical studies to investigate the possibility of using this personality dimension as a predictor for HF-rTMS driven antidepressant effects. Of note, previous HF-rTMS studies have reported persistence[63] and self-directedness[64] scores of the TCI as good predictors of antidepressant effects. Upon post-hoc analysis, we did not find such correlations between other TCI dimensions and HF-rTMS induced changes in the sgACC and this might be, since our sample consisted of healthy volunteers. Future studies in depression cohorts should address this issue.

As noted earlier, sgACC activity has been implicated in sad mood[31,32] and rumination thoughts[13] in healthy participants. We hence studied the relationship between the functional connectivity changes in sgACC and the reported negative emotional state using PANAS scale, which we administered before and after stimulation. However, we could not identify any relationship in sgACC with these measurements. Taking into consideration that the right vStr displayed robust reduction in functional connectivity during the same time window in which the sgACC decoupled from the DMN and that the NAcc is implicated in integrating emotional signals within the hubs of limbic system[37], we investigated the relationship between this region and the reported negative emotional state of subjects. Although this relationship does not reach significance, there is a correlation trend only in the

We created a new method of personalized HF-rTMS target selection utilizing individual rsfMRI and used this approach to shed new light on the DMN changes resulting from of a single, complete dose of 10 Hz rTMS stimulation in healthy subjects. Since this translational approach has been initially implemented in healthy subjects only, it is obviously not possible to comment on the clinical efficacy of stimulating at such indDLPFC targets. Our study's main goal was to delineate the underlying neural mechanisms of a full single dose of HF-rTMS protocol. Therefore, we chose sham over an alternative target selection process for direct comparison. This limits claims about how effective would the new personalization method be in the clinical practice, which will be tested in a new study. Multiple sessions of HF-rTMS might interact with brain networks differently than a single session, especially with the complexities and variations added upon by depression pathophysiology and treatment resistance. These questions remain to be further addressed in follow up studies. Another limiting aspect is that the sham condition was in fact an active sham condition, although with irrelevant current density, that did not result in functional connectivity differences. The use of a passive sham coil is unlikely to give different results from those presented here, but this should nevertheless be rigorously tested. Lastly, the correlation in self-rated emotional judgement was not statistically significant in the vStr and non-existent in the sgACC. We consider two possible explanations for that, one being that the sample size is limited in power, the other being that participants are healthy and only one session of HF-rTMS is insufficient to exhibit stronger effects. Nevertheless, new insights into the mechanisms associated with a full session of HF-rTMS can be gained from our results.

Materials and Methods

Participants.　We recruited healthy male and female subjects between the ages of 18–65 with no current or prior psychiatric disorders, evaluated by structured clinical interviews (see below). Exclusion criteria were: current/history of neurological or psychiatric disorder, recreational drug use in the past month, current/history of substance abuse or dependence, contraindications to the MRI scanner (e.g. metal parts in the body) or TMS application (e.g. epilepsy), pregnancy, history of traumatic brain injury, unwillingness to consent or to be informed of incidental findings, current use of anticonvulsant drugs, or prior TMS or ECT application in the past 8 weeks. The Ethics Committee of the University of Medical Centre Göttingen has approved the study protocol. All experiments were performed in accordance with relevant guidelines and regulations[21,65].

Study design.　We conducted a double-blind (subjects and interviewer), sham controlled, crossover study with healthy subjects. The experiments were conducted on 3 days with approximately one week between each appointment. Based on the resting state network pathophysiology of depression, we hypothesized that a single session of HF-rTMS would induce changes in the DMN[5], which would be associated with self-reported behavioural changes in affect and with stable personality trait of harm avoidance.

Day 1.　The subjects visited the lab on 3 occasions (day1, day2, and day3). On day 1 the subjects underwent a Structured Clinical Interview (SCI) consisting of Montgomery-Asberg Depression Rating Scale (MADRS), Beck Depression Inventory II (BDI II), Young Mania Rating Scale (YMRS), and Hamilton Depression Rating Scale (HAM-D). Apart from the SCI they also completed the Symptoms Checklist 90 (SCL 90), Positive and Negative Syndrome Scale (PANSS), Barratt Impulsiveness Scale (BIS), Temperament and Character Inventory (TCI), Life Orientation Test – Revised (LOT-R) and vocabulary intelligence test (MWT), and handedness questionnaire[66]. Upon being determined for inclusion/exclusion criteria, the subjects gave their written informed consent following which initial structural T1 and rsfMRI scans were recorded for our novel method of personalized target selection (see Fig. 8 for further details).

Day 2 and Day 3.　Day 2 and day 3 were at least one week apart to allow wash out of any HF-rTMS stimulation effects. At the beginning of experiment on day 2 and day 3, we asked each subject to complete the PANAS. PANAS is composed of 20 items that measure positive and negative dimensions of affect. To evaluate negative affect we used the total raw score (1–5 points for each item) from the negative dimension items[67].

After resting motor threshold (RMT) determination, we applied 10 Hz rTMS at personalized left DLPFC target, guided by an online neuronavigation system (Visor 1 software, ANT Neuro, Enschede, Netherlands) at 110% of RMT. We obtained a baseline rsfMRI scan (R0), and then delivered HF rTMS stimulation (either real or sham in a pseudo-randomized and counterbalanced way) to the subject at the pre-selected target, guided by online neuronavigation. In the event of extreme scalp discomfort, the stimulation was stopped and the subject was excluded from further experiments. Figure 8 pictorially details the study design.

Target selection.　For real and sham HF-rTMS on day 2 and day 3, we used the rsfMRI scan from day 1 to identify a personalized target in the left DLPFC for each subject, using a novel selection process as detailed in the Supplementary Information.

rTMS Stimulation.　We delivered 10 Hz rTMS using a MagVenture X100 with Mag-option and an MCF-B65 cooled butterfly coil. The stimulation parameters were as described in O'Reardon *et al.*[68]. To deliver sham stimulation, we rotated the coil by a full 180° along the handle axis of the coil.

To deliver sham stimulation, we rotated the coil by a full 180° along the handle axis of the coil such that the stimulation side of the coil faced away from the scalp and the distance between the stimulation side and the scalp was larger than 5 cm. We made measurements of the voltage induced on the "sham" side using a standard oscilloscope. The oscilloscope readings indicated that there was a very weak current strength produced by the sham side of the coil, and this negligible current did not elicit any motor responses, irrespective of how high the stimulator

Imaging acquisition and analysis. We acquired the structural (T1- and T2-weighted scans with 1-mm isotropic resolution) and functional data with a 3T MR scanner (Magnetom TRIO, Siemens Healthcare, Erlangen, Germany) using a 32-channel head coil. The gradient-echo EPI sequence had the following parameters: TR of 2.5 seconds, TE of 33 ms, 60 slices with a multiband factor of 3, FOV of 210 mm × 210 mm, 2 × 2 × 2 mm, with 10% gap between slices and anterior to posterior phase encoding. The rsfMRI data was acquired with 125 volumes in approx. 5.5 minutes, whereas the finger tapping data was acquired with 103 volumes in approx. 4.5 minutes.

rsfMRI data analysis: Using SPM12 and MATLAB (The MathWorks, Inc., Natick, MA, USA), we preprocessed the individual rsfMRI data using standard steps: slice time correction, motion correction, individual gradient echo field map unwarping, normalization, and regression of white matter, cerebrospinal fluid and motion nuisance parameters. We then temporally concatenated the data to perform group independent component analysis with FSL 5.0.7 software[69]. We identified the best fitting inde-pendent component (IC) that resembled the DMN. This IC was then back reconstructed in individual subjects' normalized rsfMRI data, r-to-z transformed and compared across the groups (Real [R0, R1, R2, R3] versus Sham [R0, R1, R2, R3]).

Finger tapping fMRI data: We preprocessed the finger tapping data using slice time correction, motion correction, gradient echo field map-based distortion correction, co-registration to the anatomical scan and smoothing with an 8 mm FWHM kernel. The onset times and durations for green dot (finger tapping) and red dot (rest period) were extracted from log files generated by Presentation software (Neurobehavioral Systems, Inc.) to create a block design of the experiment. Estimates of neural activity were computed with a general linear model (GLM) for each subject individually using SPM12. First-level contrasts were calculated for the finger tapping and rest response blocks. By contrasting these blocks, we obtained the primary motor cortex areas activated by finger tapping.

Extraction of betas (functional connectivity strengths) from IC-ACC: To evaluate potential correlations between the left DLPFC and the sgACC in the IC-ACC, we compared the personalized targets with targets based on fixed MNI coordinates, as described in a recent study[19]. We used MarsBar[70] to extract the parameter estimates (beta weights) utilizing an ROI of 2 mm radius around the personalized left DLPFC locations for real HF-rTMS stimulation, and around the standard MNI locations of the DLPFC and sgACC (as described in Tik *et al.*[19]), after transforming the coordinates into the individual anatomical spaces of our participants. To correlate functional connectivity strengths of sgACC and NAcc with the self-reported negative emotions, we utilized the same proto-col as above to extract the parameter estimates of the right NAcc using a spherical ROI of 5 mm radius centred at [(10, 12, −8) based on Harvard-Oxford Sub-cortical Atlas of FSL] and of the right sgACC using a spherical ROI of 5 mm radius centred at [(8 40−6) based on coordinates reported in Tik *et al.*[19]]. The results presented are for instances in which the beta weights were successfully extracted using MarsBar.

Statistical analysis. We used SPM12 to compare time windows of rsfMRI across real and sham conditions using a factorial design ANOVA, and only report results surviving a statistical threshold of whole brain $p < 0.05$ FWE correction for multiple testing. We used SPSS and Matlab to run two way t-tests to compare the scores from MADRS, HAM-D, YMRS, BDI II, PANAS, and VAS for real and sham stimulation sessions. Using MATLAB and R, we ran Pearson's correlation tests between functional connectivity strengths of various brain regions and relevant behavioural scales.

Data Availability

The datasets generated and analysed during the current study are not publicly available due to restrictions in the data sharing permissions obtained from study participants.

References

1. Rahnev, D. *et al.* Continuous theta burst transcranial magnetic stimulation reduces resting state connectivity between visual areas. *J. Neurophysiol.* **110**, 1811–21 (2013).
2. Ji, G.-J., Yu, F., Liao, W. & Wang, K. Dynamic aftereffects in supplementary motor network following inhibitory transcranial magnetic stimulation protocols. *Neuroimage* **149**, 285–294 (2017).
3. Halko, M. A., Farzan, F., Eldaief, M. C., Schmahmann, J. D. & Pascual-Leone, A. Intermittent theta-burst stimulation of the lateral cerebellum increases functional connectivity of the default network. *J. Neurosci.* **34**, 12049–12056 (2014).
4. Wang, J. X. *et al.* Targeted enhancement of cortical-hippocampal brain networks and associative memory. *Science (80-.).* **345**, 1054–1057 (2014).
5. Liston, C. *et al.* Default mode network mechanisms of transcranial magnetic stimulation in depression. *Biol. Psychiatry* **76**, 517–526 (2014).
6. Greicius, M. D. *et al.* Resting-State Functional Connectivity in Major Depression: Abnormally Increased Contributions from Subgenual Cingulate Cortex and Thalamus. *Biol. Psychiatry* **62**, 429–437 (2007).
7. Li, B. *et al.* A Treatment-Resistant Default Mode Subnetwork in Major Depression. *Biol. Psychiatry* **74**, 48–54 (2013).
8. Manoliu, A. *et al.* Insular dysfunction within the salience network is associated with severity of symptoms and aberrant inter-network connectivity in major depressive disorder. *Front. Hum. Neurosci.* **7**, 1–17 (2014).
9. Zhu, X. *et al.* Evidence of a dissociation pattern in resting-state default mode network connectivity in first-episode, treatment-naive major depression patients. *Biol. Psychiatry* **71**, 611–617 (2012).
10. van Tol, M.-J. *et al.* Local cortical thinning links to resting-state disconnectivity in major depressive disorder. *Psychol. Med.* **44**, 2053–2065 (2014).
11. Sheline, Y. I., Price, J. L., Yan, Z. & Mintun, M. A. Resting-state functional MRI in depression unmasks increased connectivity between networks via the dorsal nexus. *Proc. Natl. Acad. Sci. USA* **107**, 11020–5 (2010).
12. Alexopoulos, G. S. *et al.* Functional connectivity in the cognitive control network and the default mode network in late-life depression. *J. Affect. Disord.* **139**, 56–65 (2012).
13. Berman, M. G. *et al.* Depression, rumination and the default network. *Soc. Cogn. Affect. Neurosci.* **6**, 548–555 (2011).

15. Andreescu, C. *et al.* Resting state functional connectivity and treatment response in late-life depression. *Psychiatry Res. Neuroimaging* **214**, 313–321 (2013).
16. Bluhm, R. *et al.* Resting state default-mode network connectivity in early depression using a seed region-of-interest analysis: Decreased connectivity with caudate nucleus. *Psychiatry Clin. Neurosci.* **63**, 754–761 (2009).
17. Mulders, P. C. P., van Eijndhoven, P. F., Schene, A. H., Beckmann, C. F. & Tendolkar, I. Resting-state functional connectivity in major depressive disorder: A review. *Neurosci. Biobehav. Rev.* **56**, 330–344 (2015).
18. Taylor, S. F. *et al.* Changes in brain connectivity during a sham-controlled, transcranial magnetic stimulation trial for depression. *J. Affect. Disord.* **232**, 143–151 (2018).
19. Tik, M. *et al.* Towards understanding rTMS mechanism of action: Stimulation of the DLPFC causes network-specific increase in functional connectivity. *Neuroimage* **162**, 289–296 (2017).
20. Fitzgerald, P. B., Maller, J. J., Hoy, K. E., Thomson, R. & Daskalakis, Z. J. Exploring the optimal site for the localization of dorsolateral prefrontal cortex in brain stimulation experiments. *Brain Stimul.* **2**, 234–237 (2009).
21. Lefaucheur, J.-P. P. *et al.* Evidence-based guidelines on the therapeutic use of repetitive transcranial magnetic stimulation (rTMS). *Clin Neurophysiol* **125**, 2150–2206 (2014).
22. Fox, M. D., Buckner, R. L., White, M. P., Greicius, M. D. & Pascual-Leone, A. Efficacy of transcranial magnetic stimulation targets for depression is related to intrinsic functional connectivity with the subgenual cingulate. *Biol. Psychiatry* **72**, 595–603 (2012).
23. Weigand, A. *et al.* Prospective Validation That Subgenual Connectivity Predicts Antidepressant Efficacy of Transcranial Magnetic Stimulation Sites. *Biol. Psychiatry* **84**, 28–37 (2018).
24. Tavor, I. *et al.* Task-free MRI predicts individual differences in brain activity during task performance. *Science (80-.).* **352**, 216–220 (2016).
25. Yang, T. T. *et al.* Adolescent subgenual anterior cingulate activity is related to harm avoidance. *Neuroreport* **20**, 19–23 (2009).
26. Hakamata, Y., Iwase, M., Kato, T., Senda, K. & Inada, T. The Neural Correlates of Mindful Awareness: A Possible Buffering Effect on Anxiety-Related Reduction in Subgenual Anterior Cingulate Cortex Activity. *PLoS One* **8** (2013).
27. Hakamata, Y. *et al.* Gender difference in relationship between anxiety-related personality traits and cerebral brain glucose metabolism. *Psychiatry Res. - Neuroimaging* **173**, 206–211 (2009).
28. Mulder, R. T., Joyce, P. R., Frampton, C. M. A., Luty, S. E. & Sullivan, P. F. Six months of treatment for depression: outcome and predictors of the course of illness. *Am. J. Psychiatry* **163**, 95–100 (2006).
29. Abrams, K. Y. *et al.* Trait and state aspects of harm avoidance and its implication for treatment in major depressive disorder, dysthymic disorder, and depressive personality disorder. *Psychiatry Clin. Neurosci.* **58**, 240–248 (2004).
30. Chen, C.-Y., Lin, S.-H., Li, P., Huang, W.-L. & Lin, Y.-H. The Role of the Harm Avoidance Personality in Depression and Anxiety During the Medical Internship. *Medicine (Baltimore).* **94**, e389 (2015).
31. Liotti, M. *et al.* Differential limbic-cortical correlates of sadness and anxiety in healthy subjects: Implications for affective disorders. *Biol. Psychiatry* **48**, 30–42 (2000).
32. Mayberg, H. S. *et al.* Reciprocal limbic-cortical function and negative mood: converging PET findings in depression and normal sadness. *Am. J. Psychiatry* **156**, 675–82 (1999).
33. Deng, Z. D., Lisanby, S. H. & Peterchev, A. V. Electric field depth-focality tradeoff in transcranial magnetic stimulation: Simulation comparison of 50 coil designs. *Brain Stimul.* **6**, 1–13 (2013).
34. Thielscher, A., Antunes, A. & Saturnino, G. B. Field modeling for transcranial magnetic stimulation: A useful tool to understand the physiological effects of TMS? In *2015 37th Annual International Conference of the IEEE Engineering in Medicine and Biology Society (EMBC)* 222–225, https://doi.org/10.1109/EMBC.2015.7318340 (IEEE, 2015).
35. van Dijk, K. R. A. A., Sabuncu, M. R. & Buckner, R. L. The influence of head motion on intrinsic functional connectivity MRI. *Neuroimage* **59**, 431–438 (2012).
36. Schulze, L. *et al.* Number of pulses or number of sessions? An open-label study of trajectories of improvement for once-vs. twice-daily dorsomedial prefrontal rTMS in major depression. *Brain Stimul.* **11**, 327–336 (2017).
37. Floresco, S. B. The Nucleus Accumbens: An Interface Between Cognition, Emotion, and Action. *Annu. Rev. Psychol.* **66**, 25–52 (2015).
38. Bakker, N. *et al.* rTMS of the Dorsomedial Prefrontal Cortex for Major Depression: Safety, Tolerability, Effectiveness, and Outcome Predictors for 10 Hz Versus Intermittent Theta-burst Stimulation. *Brain Stimul.* **8**, 208–215 (2015).
39. Lefaucheur, J. P. *et al.* The value of navigation-guided rTMS for the treatment of depression: An illustrative case. *Neurophysiol. Clin.* **37**, 265–271 (2007).
40. Lefaucheur, J. P. Why image-guided navigation becomes essential in the practice of transcranial magnetic stimulation. *Neurophysiol. Clin.* **40**, 1–5 (2010).
41. Ruohonen, J. & Karhu, J. Navigated transcranial magnetic stimulation. *Neurophysiol. Clin. Neurophysiol.* **40**, 7–17 (2010).
42. Du, L. *et al.* Stimulated left DLPFC-nucleus accumbens functional connectivity predicts the anti-depression and anti-anxiety effects of rTMS for depression. *Transl. Psychiatry* **7**, 3 (2017).
43. Fox, M. D., Liu, H. & Pascual-Leone, A. Identification of reproducible individualized targets for treatment of depression with TMS based on intrinsic connectivity. *Neuroimage* **66**, 151–160 (2013).
44. Hamilton, J. P. *et al.* Default-Mode and Task-Positive Network Activity in Major Depressive Disorder: Implications for Adaptive and Maladaptive Rumination. *Biol. Psychiatry* **70**, 327–333 (2011).
45. Drevets, W. C. Neuroimaging studies of mood disorders. *Biol. Psychiatry* **48**, 813–829 (2000).
46. Nugent, A. C., Robinson, S. E., Coppola, R. & Zarate, C. A. Preliminary differences in resting state MEG functional connectivity pre- and post-ketamine in major depressive disorder. *Psychiatry Res. - Neuroimaging* **254**, 56–66 (2016).
47. Jacobs, R. H. *et al.* Decoupling of the amygdala to other salience network regions in adolescent-onset recurrent major depressive disorder. *Psychol. Med.* **46**, 1055–1067 (2016).
48. Connolly, C. G. *et al.* Resting-state functional connectivity of subgenual anterior cingulate cortex in depressed adolescents. *Biol. Psychiatry* **74**, 898–907 (2013).
49. Musgrove, D. R. *et al.* Impaired Bottom-Up Effective Connectivity Between Amygdala and Subgenual Anterior Cingulate Cortex in Unmedicated Adolescents with Major Depression: Results from a Dynamic Causal Modeling Analysis. *Brain Connect.* **5**, 608–619 (2015).
50. Baeken, C. *et al.* The impact of accelerated HF-rTMS on the subgenual anterior cingulate cortex in refractory unipolar major depression: Insights from 18FDG PET brain imaging. *Brain Stimul.* **8**, 808–815 (2015).
51. Nugent, A. C., Robinson, S. E., Coppola, R., Furey, M. L. & Zarate, C. A. Group differences in MEG-ICA derived resting state networks: Application to major depressive disorder. *Neuroimage* **118**, 1–12 (2015).
52. Hamani, C. *et al.* The Subcallosal Cingulate Gyrus in the Context of Major Depression. *Biol. Psychiatry* **69**, 301–308 (2011).
53. Dutta, A., McKie, S. & Deakin, J. F. W. Resting state networks in major depressive disorder. *Psychiatry Res. Neuroimaging* **224**, 139–151 (2014).
54. Herringa, R. J. *et al.* Childhood maltreatment is associated with altered fear circuitry and increased internalizing symptoms by late adolescence. *Proc. Natl. Acad. Sci. USA* **110**, 19119–24 (2013).

56. Berlim, M. T., McGirr, A., Van den Eynde, F., Fleck, M. P. & Giacobbe, P. Effectiveness and acceptability of deep brain stimulation(DBS) of the subgenual cingulate cortex for treatment resistant depression: A systematic review and exploratory meta-analysis. *Journal of affective disorders* **159**, 31–38 (2014).
57. Gong, L. *et al.* Disrupted reward circuits is associated with cognitive deficits and depression severity in major depressive disorder. *J. Psychiatr. Res.* **84**, 9–17 (2017).
58. Kampman, O. *et al.* Temperament profiles, major depression, and response to treatment with SSRIs in psychiatric outpatients. *Eur. Psychiatry* **27**, 245–249 (2012).
59. Ward, J. *et al.* Polygenic risk scores for major depressive disorder and neuroticism as predictors of antidepressant response: Meta-analysis of three treatment cohorts. *PLoS One* **13**, e0203896 (2018).
60. De Fruyt, F., Van De Wiele, L. & Van Heeringen, C. Cloninger's psychobiological model of temperament and character and the five-factor model of personality. *Pers. Individ. Dif.* **29**, 441–452 (2000).
61. Quilty, L. C. *et al.* Dimensional personality traits and treatment outcome in patients with major depressive disorder. *J. Affect. Disord.* **108**, 241–250 (2008).
62. Kim, S.-Y. *et al.* Influences of the Big Five personality traits on the treatment response and longitudinal course of depression in patients with acute coronary syndrome: A randomised controlled trial. *J. Affect. Disord.* **203**, 38–45 (2016).
63. Siddiqi, S. H., Chockalingam, R., Cloninger, C. R., Lenze, E. J. & Cristancho, P. Use of the temperament and character inventory to predict response to repetitive transcranial magnetic stimulation for major depression. *J. Psychiatr. Pract.* **22**, 193–202 (2016).
64. Baeken, C. *et al.* Self-directedness: An indicator for clinical response to the HF-rTMS treatment in refractory melancholic depression. *Psychiatry Res.* **220**, 269–274 (2014).
65. Rossi, S. *et al.* Safety, ethical considerations, and application guidelines for the use of transcranial magnetic stimulation in clinical practice and research. *Clin. Neurophysiol.* **120**, 2008–2039 (2009).
66. Oldfield, R. C. The assessment and analysis of handedness: The Edinburgh inventory. *Neuropsychologia* **9**, 97–113 (1971).
67. Liu, Q., Zhou, R., Chen, S. & Tan, C. Effects of Head-Down Bed Rest on the Executive Functions and Emotional Response. *PLoS One* **7** (2012).
68. O'Reardon, J. P. *et al.* Efficacy and Safety of Transcranial Magnetic Stimulation in the Acute Treatment of Major Depression: A Multisite Randomized Controlled Trial. *Biol. Psychiatry* **62**, 1208–1216 (2007).
69. Jenkinson, M., Beckmann, C. F., Behrens, T. E. J., Woolrich, M. W. & Smith, S. M. FSL. *Neuroimage* **62**, 782–790 (2012).
70. Brett, M., Anton, J. -L., Valbregue, R. & Poline, J. -B. Region of interest analysis using an SPM toolbox. *In NeuroImage* **16**, 769–1198 (Academic Press, 2002).

5 Default mode network alterations after intermittent theta burst stimulation in healthy subjects

High frequency 10 Hz rTMS became the first rTMS protocol to receive FDA approval in 2008 to treat depression and has since frequently been used in clinical settings. The accumulated evidence supports the relatively shorter iTBS protocol (a quarter of the time or less per session) as an efficient antidepressant treatment [96, 100, 111, 258]. ITBS protocol is now established with FDA approval since 2018 as a non-inferior antidepressant compared to conventional 10 Hz rTMS protocol [88]. In this study, we aimed to uncover connectivity effects of a single session of a prolonged iTBS protocol (1800 pulses) in healthy subjects.

Akin to 10 Hz rTMS, brain connectivity changes (especially of DMN) resultant from iTBS delivered at the DLPFC remain unexplored. As with other rTMS protocols, several factors, including natural variation in anatomy and functional connectivity, contribute to inter-individual variability. Comparable to our last experiment, we address this variability using the target selection method presented and validated previously (chapter 4). Using the personalized left DLPFC target sites, we investigate the effects of iTBS on the DMN. We use a double-blind, crossover, and sham- controlled study design identical to that employed previously (chapter 4) and use the individual rsfMRI data to locate the personalized targets. We applied a single session of iTBS (1800 pulses) at personalized left DLPFC sites followed by analysis of the DMN during three time-windows after stimulation. The time lapse of effects from iTBS at left DLPFC have not been reported before. We suppose depicting the DMN effects from one session of iTBS stimulation in healthy subjects will aid understanding the relationship between the stimulation site and the sgACC across time, thus, promoting tailored and more efficient iTBS interventions in the future.

Depression is characterized by a hyperconnected DMN [146, 250] and "normalization" of this hyperconnectivity suggests efficient treatment for depression [159]. As iTBS has been shown to be an effective antidepressant treatment [88, 142, 263–266, 144, 146, 148, 158, 259–262], we expect that iTBS would influence the DMN by reducing its functional connectivity of healthy subjects. This hypothesis is naturally derived from the assumption that the neural mechanism of multiple sessions of iTBS in patients would be a simple accumulation of the effect observed after a single session in healthy subjects. This assumption is justified, as we previously reported (chapter 4) certain congruences between changes seen after a session of 10 Hz rTMS in healthy subjects and those reported in patients with depression. Based on the results from the previous chapter, we expected to see maximum effects of a single session of iTBS after approximately 30 min post stimulation. Lastly, as seen in the literature [175, 176, 251–253, 267] and in our 10 Hz rTMS results (chapter 4), we explored the role of HA in relation to the modulation of DMN connectivity after stimulation. Based on a negative relationship between induced changes in sgACC and HA, we hypothesize that a similar negative relationship would exist between iTBS induced changes and HA scores. Further details on the used methodology, the outcomes of the experiment and a corresponding discussion of the results are presented below.

56

My contributions pertain to data collection and analysis and writing the original and revised drafts of the manuscript. The article was published under [Creative Commons Attribution 4.0 International License](#).

Singh et al. *Translational Psychiatry* (2020)10:75
https://doi.org/10.1038/s41398-020-0754-5

Translational Psychiatry

Default mode network alterations after intermittent theta burst stimulation in healthy subjects

Aditya Singh[1], Tracy Erwin-Grabner[1], Grant Sutcliffe[1], Walter Paulus[2], Peter Dechent[3], Andrea Antal[2] and Roberto Goya-Maldonado[1]

Introduction

The large variability of responses to the FDA-approved 10 Hz repetitive transcranial magnetic stimulation (rTMS) protocol for the treatment of depression has led to a world-wide demand for better techniques or improved protocols. The non-inferior antidepressant efficacy of the 3 min/session theta burst protocol[1] compared to 37.5 min/sessions of conventional 10 Hz rTMS protocol has played a role in increasing the use of the theta burst protocol for antidepressant treatment[2–5]. Nevertheless,

brain connectivity changes underlying the effects of intermittent theta burst stimulation (iTBS) delivered at the left dorsolateral prefrontal cortex (DLPFC) remain unexplored. Many factors contribute to inter-individual variability, including natural variation in anatomy and functional connectivity. Here we use a previously validated target selection method to improve precision of coil localization and investigated the effects of iTBS on the relevant brain networks that best cover the left DLPFC and the anterior cingulate cortex (ACC).

The TBS protocol was developed to mimic rodent[6,7] and human hippocampal activity[8], where a combination of gamma-frequency spike patterns superimposed on theta rhythms[9] was found. It involves application of a burst of three TMS pulses every 20 ms (50 Hz), which is repeated five times per second (5 Hz)[10,11]. When delivered continuously (continuous TBS–cTBS) for 40 s, it results

---

Correspondence: Roberto Goya-Maldonado (roberto.goya@med.uni-goettingen.de)
[1]Laboratory of Systems Neuroscience and Imaging in Psychiatry, Department of Psychiatry and Psychotherapy of the University Medical Center Göttingen, Göttingen, Germany
[2]Department of Clinical Neurophysiology of the University Medical Center Göttingenn, Göttingen, Germany
Full list of author information is available at the end of the article.

in reduced corticospinal excitability, while when administered in an intermittent fashion (iTBS) it results in increased corticospinal excitability[9]. Studies of TBS stimulation on motor cortex have shown plasticity changes beyond the duration of stimulation typically lasting in the range of 30 min[11,12].

Beyond local effects under the stimulation coil, plasticity changes in brain's altered functional connectivity away from the stimulation point, e.g. the DLPFC[13], are likely relevant to the treatment of depression, which has been associated with aberrant brain functional connectivity[14]. The default-mode network (DMN), consisting of the medial prefrontal cortex, posterior cingulate cortex, and areas of posterior parietal cortex[15]. Using 10 Hz rTMS as antidepressant treatment, a study has recently replicated the prediction of symptomatic alleviation in depression when aberrant sgACC connectivity with the DMN is decreased, which happened in responders but not in non-responders[18]. Furthermore, such effects over networks in healthy subjects have been shown in our previous work using 10 Hz rTMS[19]. Already after a single session of 10 Hz rTMS (3000 pulses), significant reduction in the connectivity between the sgACC and the DMN was evidenced.

Given the central involvement of the DMN in the pathophysiology of depression and the importance of a shorter protocol such as iTBS for reducing symptoms[15,20–31], here we aimed to uncover connectivity effects of a single session of a prolonged iTBS protocol (1800 pulses) in healthy subjects. ITBS is therapeutically beneficial for depression[1], which is characterized by a hyperconnected DMN[15,16], and treatment for depression is accompanied by normalization of this hyperconnectivity[17]. Assuming the therapeutic mechanism of iTBS shares the same mechanism of action in healthy subjects, we hypothesize that iTBS would act by reducing the functional connectivity of DMN.

We applied a single session of iTBS at left DLPFC sites and analyzed the DMN during three time-windows after stimulation in a double-blind, crossover, and sham-controlled study. To the best of our knowledge time lapse of effects from iTBS at left DLPFC have not been previously reported. We expect that depicting the acute effects after one session of iTBS stimulation in healthy subjects will be crucial to understand the relationship between the stimulation site and the sgACC across time, up to ~50 min after stimulation[32]. Further comprehension of these network dynamics could help to inform and promote more efficient and tailored iTBS interventions in the future. Considering our previous results from a single session of 10 Hz rTMS, we expected to see maximum effects of a single session of iTBS after approximately 30 min. We hence acquired a resting state functional MRI

exploratory basis, we also wanted to investigate potential changes during other time windows. Particularly, to document iTBS induced changes before and after the expected 30-min mark, we acquired one rsfMRI from 10 to 15 min (earliest possible time point after stimulation) and another from 45 to 50 min after stimulation.

Lastly, harm avoidance (HA) of the Temperament and Character Inventory (TCI) pertains to the heritable tendency of individuals to respond more harshly to aversive cues, punishment and non-reward[33]. Previous works show a relationship between HA and activity[34–37] or connectivity[19] in sgACC in healthy samples. In patients with depression, HA has been shown to be associated with response[38] and non-response[39] to treatment. As we employ a single session of clinically relevant iTBS protocol in healthy subjects, we propose that reduction of DMN connectivity after stimulation would have a relationship to HA. Based on our previously reported[19] negative relationship between 10 Hz rTMS induced changes in sgACC and HA, we hypothesize that a similar negative relationship would exist between iTBS induced changes and HA scores.

Materials and methods

Participants

Healthy subjects between the ages of 18–65 were enrolled in the study. We evaluated the subjects with structured clinical interviews and ruled out current or prior neuropsychiatric disorders and contraindications to rTMS and/or MRI. We performed the experiments in agreement with relevant guidelines and regulations[40,41]. The Ethics Committee of the University of Medical Center Göttingen approved the study protocol and subjects provided their informed consent before investigation.

Study design

The study reported here with healthy subjects is a sham-controlled, double-blind (subject and interviewer were unaware of the stimulation condition), crossover study with real and sham iTBS delivered in a counterbalanced and pseudo-randomized fashion. We conducted the experiments over three sessions (each session on a different day, Fig. 1) with each session separated by at least one week.

Session 1

In session 1, the interviewer administered a Structured Clinical Interview (SCI) consisting of the Beck Depression Inventory II (BDI II), Montgomery-Asberg Depression Rating Scale (MADRS), Hamilton Depression Rating Scale (HAM-D) and Young Mania Rating Scale (YMRS). In addition to the SCI, to further establish the mental health

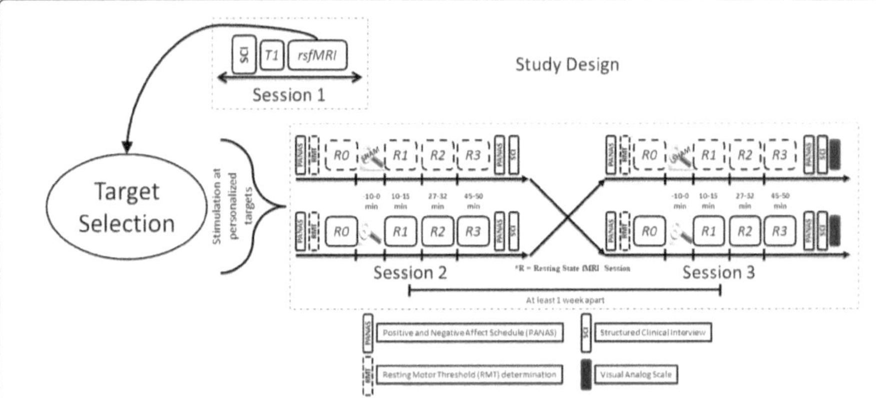

Fig. 1 A schematic representation of the study design. In session 1, we obtained the informed consent and collected the information (see main text for details) from Structured Clinical Interviews (SCI). After this we acquired a structural (T1-weighted) and functional (rsfMRI) images. During rsfMRI, the subjects were instructed to fixate at a "+" and mind wander, while their open eyes were monitored by eye tracking. Personalized targets were found using each subject's rsfMRI as has been described elsewhere[19]. Using online neuronavigation, we delivered real or sham iTBS, in a counterbalanced and pseudo-randomized fashion, at 80% of resting motor threshold. Baseline and three post-iTBS rsfMRI scans were acquired. The subjects also completed the Positive and Negative Affect Schedule (PANAS) both before and after the sessions, and a visual analog scale (VAS) for perceived effects of iTBS on mental state and scalp sensation at the end of the investigation.

the Symptoms Checklist 90-revised (SCL 90-R), Temperament and Character Inventory (TCI), Positive and Negative Syndrome Scale (PANSS), Life Orientation Test – Revised (LOT-R), Barratt Impulsiveness Scale (BIS), a handedness questionnaire[42] and a vocabulary-based intelligence test (MWT). Next, we acquired structural T1-weighted MRI and resting state functional MRI (rsfMRI) scans for our method of target selection (see Fig. 1 for further details). The process of personalized left DLPFC target selection has been described previously[19].

Session 2 and Session 3

To allow washout of any potential iTBS effects, session 2 and session 3 were separated by at least a week. After determining the resting motor threshold (RMT), we applied iTBS, at 80% RMT. We navigated to the individual left DLPFC target using an online neuronavigation system (Visor 1 software, ANT Neuro, Enschede, Netherlands). We obtained a pre-iTBS (baseline) rsfMRI scan (R0) followed by three post-iTBS rsfMRI scans (R1, R2, R3). Subjects completed the Positive and Negative Affect Schedule (PANAS[43]) both before and after the experiment on session 2 and session 3. This allowed us to follow any short-term changes in the subjects' mood potentially influenced by iTBS. Figure 1 schematically shows the study design.

rTMS protocol

We delivered iTBS using a MagVenture X100 with coil at the targets selected using each individual subject's rsfMRI (see ref. [19]). We used stimulation parameters from Li C-T et al.[5] (3 pulses burst at 50 Hz delivered at 5 Hz for 2 s with an 8 s inter train interval, total 60 trains delivered during 9 min 30 s). For the sham condition, we rotated the coil by 180° along the handle axis as described elsewhere[19]. We did not employ the usual method of rotating the coil by 90°, as a complete 180° rotation allowed us to make the sham condition look as similar as possible to the real condition, for more effective blinding of subjects.

Image acquisition

We collected functional data and, in between the rsfMRI scans, structural (T1- and T2-weighted scans with 1-mm isotropic resolution) data with a 3T MR scanner (Magnetom TIM TRIO, Siemens Healthcare, Erlangen, Germany) using a 32-channel head coil. The T2*-weighted multi-band gradient echo echo-planar imaging sequence provided by the Center for Magnetic Resonance Research of the University of Minnesota[44,45] had the following parameters: repetition time of 2.5 s, echo time of 33 ms, flip angle of 70°, 60 axial slices with a multi-band factor of 3, $2 \times 2 \times 2\,mm^3$, FOV of 210 mm, with 10% gap between slices and posterior to anterior phase encoding. The rsfMRI data were acquired with 125 volumes in approx. 5 min. The gradient echo field map was acquired with repetition time of 603 ms, echo times of 4.92 ms (TE 1) and 7.38 ms (TE 2), flip angle of

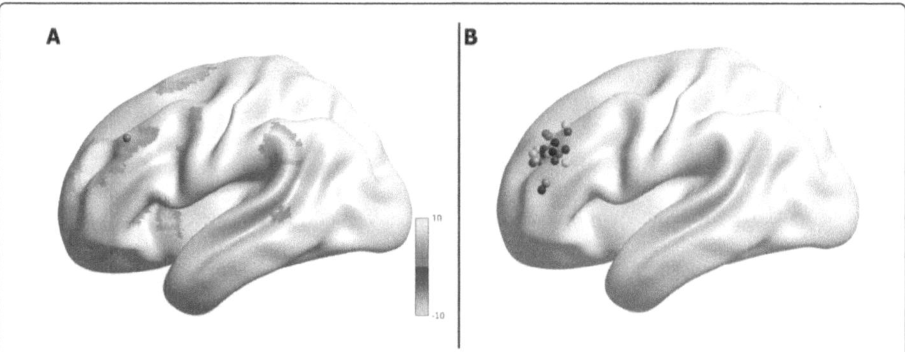

Fig. 2 Personalized left DLPFC sites. a An example of IC-DLPFC (warm colors) for a single subject from which the parameter estimates of personalized stimulation site (blue sphere) were extracted. **b** Personalized left DLPFC stimulation sites of all subjects from which parameter estimates were extracted.

gap between slices and anterior to posterior phase encoding.

Imaging data analysis

We preprocessed the rsfMRI data, using SPM12 and MATLAB (The MathWorks, Inc., Natick, MA, USA), to execute the following state-of-the-art steps: slice time correction, motion correction, gradient echo field map unwarping, normalization, and regression of motion nuisance para-meters, cerebrospinal fluid and white matter. Following this, we temporally concatenated the data for group independent component analysis (ICA) with FSL 5.0.7 software[46]. We visually identified the independent component (IC) that best resembled the DMN and another IC that best covered the left DLPFC (IC-DLPFC). We back reconstructed this IC representing the DMN in the normalized rsfMRI data of individual subjects, r-to-z transformed and compared across the groups using a factorial design ANOVA (Real [R0, R1, R2, R3] versus Sham [R0, R1, R2, R3]).

Extraction of parameter estimates (functional connectivity strengths)

We used MarsBar[47] to extract the parameter estimates (beta weights) of the rostral anterior cingulate cortex (rACC; 5 mm radius sphere) and subgenual anterior cingulate cortex (sgACC; 5 mm radius sphere) centered on independent coordinates from a meta-analysis of functional large-scale networks in depression[48] and our previous work on 10 Hz rTMS effects on DMN[19], respectively. The parameter estimates for the individual left DLPFC sites were extracted using 2 mm radius sphere region of interest (ROI) centered around the targets, in

line with our previous work[19]. Figure 2a highlights an example subject showing the IC-DLPFC (in warm color), the network from which the parameter estimates using an individual left DLPFC target ROI (blue sphere) is extracted. Figure 2b shows all the ROIs that were used for parameter estimate extraction.

Statistical analysis

Previous data on twenty-three subjects have shown adequate power to detect the differences in DMN after rTMS[19]. Using a factorial design ANOVA in SPM12 we compared the time windows of rsfMRI across real and sham conditions, and report results surviving a statistical threshold of $p < 0.05$ FWE whole-brain corrected for multiple testing. We ran Pearson's correlation tests between rACC functional connectivity strengths and the HA domain of the TCI using MATLAB. We used R to run two-way t-tests to compare the scores from YMRS, HAM-D, MADRS, PANAS, VAS, and BDI II for real and sham stimulation sessions.

Results

Twenty-nine healthy subjects (11 females, mean age of 28 ± 8 years) signed up for the study. Two subjects (both females) were dropped from the study due to failure to locate their personalized left DLPFC target and one subject (male) dropped out of study due to discomfort from stimulation. Thus, 26 subjects were included in final analysis, none of whom reported any adverse effects during or after stimulation.

Functional connectivity changes after real stimulation

After a full single session of iTBS (1800 pulses) we observed reduced functional connectivity of the rACC

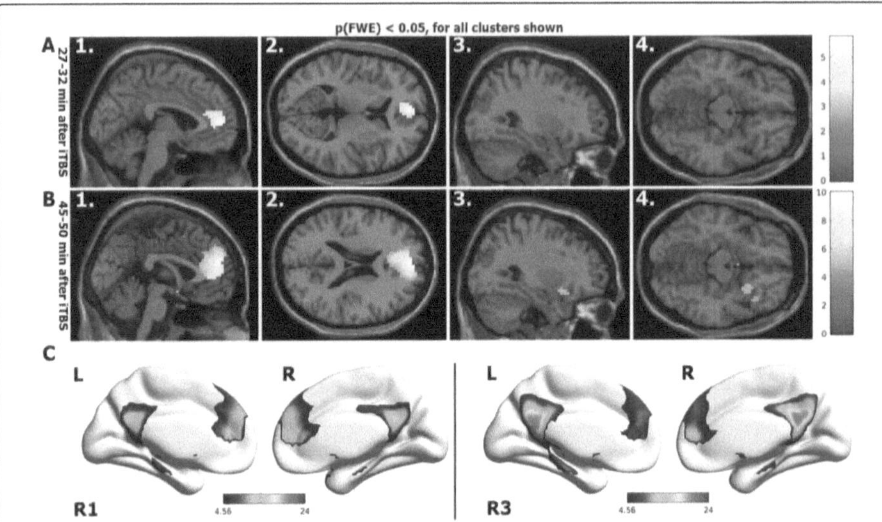

Fig. 3 Functional connectivity results. Regions that show reduced functional connectivity to DMN after stimulation (real-sham condition, whole-brain corrected $p_{FWE} < 0.05$): (**a**1-2) About 27 min after iTBS, the rACC and dACC disengage from the DMN. (**b**1-2) About 45 min after stimulation, the functional connectivity has further reduced, extending to the mPFC and (**b**3-4) the right AI. **c** DMN during R1 rsfMRI and R3 rsfMRI session after real stimulation.

rsfMRI session (27–32 min post-stimulation) when compared to R1 rsfMRI session (10–15 min post-stimulation) (Fig. 3 [A1-A2]). Even more interesting was the effect on the functional connectivity of DMN during the R3 rsfMRI (45–50 min post-stimulation), which increased in spatial extent. During R3, the area of significantly reduced functional connectivity of the DMN spread to include the medial prefrontal cortex (mPFC) and frontal poles, as seen in Fig. 3 [B1-B2]. Additionally, the right anterior insula (AI) showed decreased functional connectivity to the DMN during R3 rsfMRI (Fig. 3 [B3-B4]). These findings were not seen in the sham condition. Changes in clinical scales were neither expected nor identified. Also, it is important to note that when comparing the DMN only across real iTBS rsfMRI sessions without sham correction, we see the same regions decoupling from the DMN (Supplementary Fig. 1), except by smaller mPFC and larger right AI blobs in the R2 rsfMRI. In this case, the decoupling of the right AI is more pronounced, showing significantly reduced functional connectivity even during the R2 rsfMRI.

Functional connectivity changes in the left DLPFC and the rACC along time

To have a better understanding of the effects of iTBS, real condition: the stimulated left DLPFC site and the rACC. We used a spherical ROI of 2 mm radius centered at the left DLPFC target to extract its parameter estimates from the IC-DLPFC (see methods for definition of IC-DLPFC). We extracted the parameter estimates of the rACC from the DMN. Following an earlier study of the ACC with a 10 Hz protocol[19], a spherical 5 mm radius ROI was used with coordinates obtained from an independent meta-analysis[48]. The plot (Fig. 4) shows that the DMN functional connectivity of the rACC increases from the R0 to the R1 rsfMRI window. Subsequently, a functional connectivity decrease in the rACC from R1 to R2 is sustained until R3. A statistically insignificant increase in the IC-DLPFC functional connectivity of the left DLPFC from R0 to R1 is also seen. This functional connectivity returns to a value close to baseline during R2 and increases during R3 rsfMRI. The green dashed line represents the correlation coefficients between the parameter estimates of the sgACC and the left DLPFC. It shows that as the effect of iTBS becomes more prominent, the correlation between these regions goes from negative to more and more positive, not returning to baseline within 50 min after iTBS. We explored the changes in functional connectivity of these regions for sham condition (Supplementary Fig. 2) and observed minor changes and 0.02) and in the correlation coefficient (between 0 and 0.17). However, the functional connectivity fluctuates around the baseline during all rsfMRI sessions.

Harm avoidance—a predictor of iTBS response?

The mean and SD (11.625 ± 6.42) of the sample are representative of a healthy population[49,50]. A chi-square test for normality indicates that the HA is normally distributed (*p*-value = 0.962). Our previous work has iden-

tified a negative relationship between HA scores and the changes induced by 10 Hz rTMS in the right sgACC during R2 rsfMRI compared to R1 rsfMRI[19]. We hence explored if such a relationship existed also between the HA scores of subjects in the current study and the observed decrease in the functional connectivity of rACC during R2 rsfMRI compared to R1 rsfMRI. We identify a positive correlation between the HA measure and the decrease in functional connectivity of the rACC, only after real stimulation (*r* = 0.6052, *p* value = 0.013) but not after sham stimulation (*r* = −0.1233, *p* value = 0.6491). This indicates that the higher the HA score of the subjects the more they showed a decrease in their rACC functional connectivity to DMN (Fig. 5).

Discussion

In this double-blind, sham-controlled study, we have determined for the first time the connectivity changes of the DMN in the healthy brain for up to 50 min after iTBS (1800 pulses protocol). As expected, after left DLPFC stimulation (Fig. 2b) we see a decrease in the functional connectivity of the DMN, mainly with the rACC and dACC during the R2 rsfMRI window (Fig. 3 A1–A2, about 27–32 min after stimulation). This decrease is sustained in the dACC and additionally extends to the mPFC and right AI during the R3 rsfMRI window (Fig. 3 B1–B4, about 45–50 min after stimulation). In agreement with the literature[18,51], we see at baseline a negative correlation between the parameter estimates of sgACC and the left DLPFC (Fig. 4, green diamond at ~20 min before iTBS). As the functional connectivity changes in both left DLPFC and rACC within their own networks (Fig. 4, red and blue curves), the negative correlation between sgACC and left DLPFC becomes progressively positive (Fig. 4, green dashed curve). Finally, we observe a positive correlation between the HA score and the connectivity changes

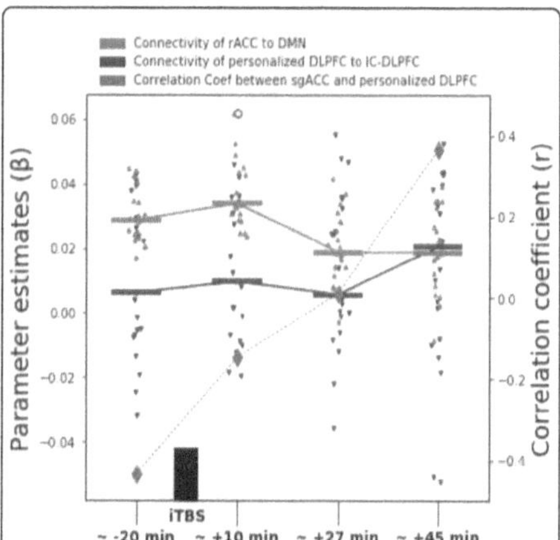

Fig. 4 Parameter estimates of left DLPFC and rACC. Left axis shows the parameter estimates of left DLPFC (blue) and rACC (red) of IC-DLPFC and DMN, respectively. Dots represent the individual values and horizontal lines depict the median of the parameter estimates for the respective rsfMRI window. The right axis plots the correlation coefficients between the DLPFC and the rACC, showing that the effect of a single session of iTBS progressively changes the correlation between these ROIs from negative to positive.

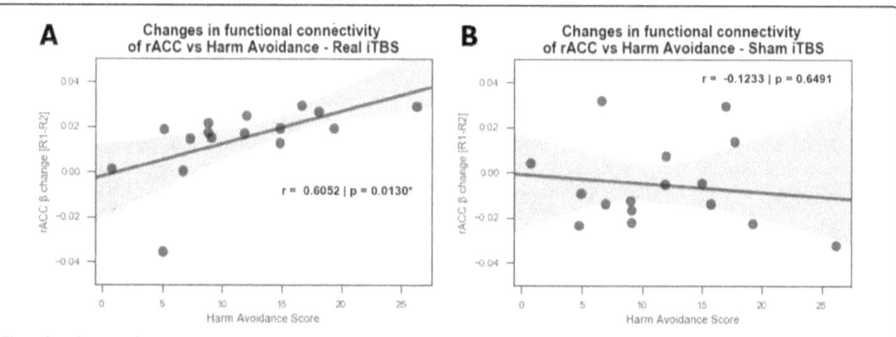

Fig. 5 Correlation to harm avoidance scores. Correlation between the HA score and the changes observed in rACC functional connectivity during R2 rsfMRI compared to R1 rsfMRI for (a) real and (b) sham conditions. A significant positive correlation is observed after real stimulation only.

observed with the DMN in the rACC (Fig. 5a), which implies that this measure can possibly predict the magnitude of functional connectivity changes induced by iTBS in the rACC.

A dynamic system known as the triple network model has been suggested to explain the fast adaptive qualities of the brain[52,53]. According to the triple network model, a task positive network corresponding to the central-executive network (CEN) is active when the brain is engaged in cognitive tasks or allocating attention to external stimuli[22,54]. The DMN (aka task negative network) is active antagonistically to the task positive network, when resources are internally allocated during introspective thoughts or autobiographical memories[55]. A dynamic interplay between the task positive and task negative network is required to quickly reallocate resources towards internal or external stimuli according to immediate demands. It has been shown that a "circuit breaker" role is played by the salience network (SN)[53], with the dACC and right AI as the main network nodes along with rACC involved with affective processing[33,56]. In this work we identified these regions as being decoupled from the DMN in healthy subjects after a single session of a prolonged iTBS protocol (1800 pulses) (Fig. 3). Although the changes evidenced here may not directly translate to the context of psychopathology, it was intriguing to see these nodes as part of our results. The AI is considered to be the essential hub of the SN because it mediates the information flow across the brain to different networks and switches between central-executive and DMNs[52,53,57].

In depression, the SN shows aberrant functional connectivity to the DMN and CEN[22,31,58]. One of the network-based hypotheses of depression conjectures that the increased interaction between the SN and the DMN results in pathologically increased allocation of resources to negative information about the self, e.g. ruminative thoughts[27,59]. Considering the proven efficacy of iTBS for treatment of depression and speculating that the effects seen in healthy participants would extend to patients, the mechanism by which iTBS may initially influence the symptomatology of depression could be by "normalizing" the pathologically increased interaction between the SN and the DMN. In line with this reasoning, Iwabuchi et al.[60] has shown in patients with depression that fronto-insular and SN connectivity interactions correlated positively with HAM-D score change at the end of a 4-week iTBS protocol. They have also described that better clinical outcomes are associated with reduced connectivity between dorsomedial prefrontal cortex (DMPFC) and bilateral insula[58]. Their results in conjunction with ours highlight the importance of investigating the AI and dACC as SN nodes

Using a different approach, Baeken et al.[61] have shown that the functional connectivity of the sgACC and medial orbitofrontal cortex (mOFC) is increased during accelerated iTBS in depression patients. Their results stem from seed-based analysis of the sgACC after iTBS. One possible reason for the discrepancy between their and our results may be the method of analysis, as seed-based analysis focuses on the functional connectivity of a predefined ROI while ICA allows exploration of functional connectivity changes of the whole brain without having to pre-define a ROI. In contrast to our previous work that identified the sgACC as the main region decoupled from the DMN after a single session of personalized 10 Hz rTMS[19], the strongest changes in connectivity after iTBS are not with the sgACC, but rather the rACC and dACC as mentioned above. However, due to the relevance of the sgACC, we further explored the beta weights from this region and evaluated its relation to the left DLPFC at baseline and up to 50-min after stimulation. We evidenced a shift of correlation between these regions from negative to positive within the observation time (Fig. 4, green dashed curve). This suggests the participation of the sgACC in the effects driven by iTBS, even though it is not directly engaged by it. The striking similarity between the red curves seen in the rACC after iTBS (Fig. 4) and in the sgACC after 10 Hz TMS (Fig. 5 in Singh et al.[19]) suggests that sgACC is rather the first target after 10 Hz rTMS (under standard dose of 3000 pulses). Another important aspect that might have contributed to differences between our and the results of Baeken et al.[61] is that we stimulated functionally relevant sites within the left DLPFC, as opposed to their structural selection of stimulation sites. Of course, the most profound difference is that our study closely evaluated connectivity changes after one session of iTBS in healthy subjects, whereas Baeken et al.[61] evaluated patients with depression after 20 stimulation sessions. It must be considered that the complexities associated with the underlying pathophysiology of depression could have contributed to differences in how iTBS interacts with brain regions and networks. Our results shed light on other relevant regions that respond to a single session of iTBS in the healthy brain. Future work examining brain networks in patients before and after 20 iTBS treatment sessions would likely close these knowledge gaps.

By rounding up results from our previous work[19] and current study, 10 Hz rTMS disengages the anterior nodes of DMN and thus affects sgACC and DMN connectivity. 10 Hz rTMS could hence reduce sadness, rumination and self-directed thought processes. In fact, we have seen a trend of reduced negative affect after 10 Hz rTMS in healthy subjects. ITBS, however, acts via a different brain network. It does not disengage DMN with itself directly but reduces the communication between nodes of SN and would engage less in affective content of external information due to reduced communication between SN and DMN. However, this needs more precise testing in future studies.

We also evidenced a positive correlation between the HA score on the TCI and changes in the functional connectivity of the rACC and the DMN (Fig. 5a). This indicates that the higher the subjects scored on the HA domain, the stronger the reduction in observed functional connectivity. This correlation indicates that it might be possible to utilize HA to predict the extent of DMN-rACC coupling changes induced by iTBS. Interestingly, the correlation between connectivity changes and HA scores replicates the time window in which this was seen in an independent sample using 10 Hz rTMS[19], although in opposite direction and in a different brain region, the sgACC.

The opposing results likely stem from the fact that 10 Hz rTMS and iTBS involve different brain networks in their action as discussed above. Previous works have shown positive relation between HA and rumination[62,63], higher levels of which are associated with lower intra anterior DMN functional connectivity[64]. Huggins et al.[65] have shown that HA correlates with the strength of anticorrelation between an SN node and DMN. This implies healthy subjects with higher HA would have lower anterior DMN functional connectivity and higher SN-DMN connectivity. Consequently, subjects with higher HA would presumably have higher rumination and greater propensity for negative valence stimuli, while subjects with lower HA would display the opposite. In this line, higher anterior DMN connectivity related to lower HA would respond more pronouncedly to 10 Hz rTMS as it directly disengages the functional connectivity between sgACC and DMN[19]. Conversely, higher SN-DMN connectivity related to higher HA would respond more conspicuously to iTBS as it disengages the nodes of SN from DMN. Following this reasoning, we propose that high HA subjects would respond oppositely to 10 Hz rTMS and iTBS.

Thus, healthy subjects with higher rumination, and anxiety and vigilance towards external stimuli (high HA) are more likely to respond to iTBS via disengagement of SN nodes from DMN. While those with lower disposition for rumination, anxiety, and vigilance towards external stimuli (low HA) are more likely to respond to 10 Hz rTMS, which directly disengages higher functional connectivity between sgACC and DMN. Considering the opposite relationship between HA and DMN effects from 10 Hz rTMS and iTBS, we speculate that HA scores may facilitate identification of subjects who will present stronger DMN changes in response to rTMS protocols. However, given clinically depressed population have a

increased rumination[27] and self-directed thoughts, such predictive use of HA requires rigorous testing. We speculate that patients with depression having HA greater than the depressed population's average might more favorably respond to iTBS while those with HA lower than the depressed population's average would respond better to 10 Hz rTMS. If such results hold true for clinical population receiving multiple sessions of rTMS, then HA could be used to determine beforehand who would benefit most from one stimulation protocol or the other. We hope future research in precision medicine will investigate this aspect, considering the direct clinical application and potential relevance to improving treatment response.

There are limitations to our study. For ethical reasons we applied a single session of 1800 pulses iTBS for uncovering its effect on the DMN in healthy brains, since applying 20 sessions of iTBS as is done in patients[66] would not be prudent. Also, the fact that the non-stimulation side of an MCF-B65 coil can have non-zero current implies that sham condition was not completely passive. It is however unlikely that sham condition has biased our main results, considering that (a) connectivity changes from sham stimulation do not show any changes analogous to effects from real condition; and (b) the same specific nodes display connectivity changes after real stimulation without sham comparison (Supplementary Fig. 1). The diseased state of the brain, e.g. in depressive state, is also likely to influence interactions between brain networks in response to multiple sessions of iTBS. Therefore, assumptions based on healthy samples must be made cautiously. Finally, we did not expect to observe any significant neural effects from iTBS beyond 50 min after stimulation, however our results indicate that iTBS effects are strongest in the rsfMRI scan from 45 to 50 min after stimulation. This information points towards the fact that iTBS effects likely last beyond the time window of our study and should be further examined in future studies.

In conclusion, by means of a double-blind, sham-controlled crossover study involving healthy subjects, we show that a single session of iTBS results in decoupling of the rostral/dorsal ACC, followed by the mPFC and the right AI, with the DMN. The interaction between the sites of stimulation at the left DLPFC and the sgACC shows a progressive shift from negative to positive correlation. Lastly, connectivity changes in the rACC induced by a single real session of iTBS in the healthy brain positively correlated with the HA score on the TCI scale.

References

1. Blumberger, D. M. et al. Effectiveness of theta burst versus high-frequency repetitive transcranial magnetic stimulation in patients with depression (THREE-D): a randomised non-inferiority trial. *Lancet* **391**, 1683–1692 (2018).

2. Bakker, N. et al. rTMS of the dorsomedial prefrontal cortex for major depression: safety, tolerability, effectiveness, and outcome predictors for 10 Hz versus intermittent theta-burst stimulation. *Brain Stimul.* **8**, 208–215 (2015).

3. Duprat, R. et al. Accelerated intermittent theta burst stimulation treatment in medication-resistant major depression: a fast road to remission? *J. Affect. Disord.* **200**, 6–14 (2016).

4. Bulteau, S. et al. Efficacy of intermittent Theta Burst Stimulation (iTBS) and 10-Hz high-frequency repetitive transcranial magnetic stimulation (rTMS) in treatment-resistant unipolar depression: Study protocol for a randomised controlled trial. *Trials* **18**, 17 (2017).

5. Li, C.-T. et al. Efficacy of prefrontal theta-burst stimulation in refractory depression: a randomized sham-controlled study. *Brain* **137**, 2088–2098 (2014).

6. Vanderwolf, C. Hippocampal electrical activity and voluntary movement in the rat. *Electroencephalogr. Clin. Neurophysiol.* **26**, 407–418 (1969).

7. O'Keefe, J. & Recce, M. L. Phase relationship between hippocampal place units and the EEG theta rhythm. *Hippocampus* **3**, 317–330 (1993).

8. Brazier, M. A. Studies of the EEG activity of limbic structures in man. *Electroencephalogr. Clin. Neurophysiol.* **25**, 309–318 (1968).

9. Chung, S. W. et al. The effects of individualised intermittent theta burst stimulation in the prefrontal cortex: A TMS-EEG study. *Hum. Brain Mapp.* **40**, 608–627 (2019).

10. Oberman, L., Edwards, D., Eldaief, M. & Pascual-Leone, A. Safety of theta burst transcranial magnetic stimulation: a systematic review of the literature. *J. Clin. Neurophysiol.* **28**, 67–74 (2011).

11. Huang, Y. Z., Edwards, M. J., Rounis, E., Bhatia, K. P. & Rothwell, J. C. Theta burst stimulation of the human motor cortex. *Neuron* **45**, 201–206 (2005).

12. Conte, A. et al. Theta-burst stimulation-induced plasticity over primary somatosensory cortex changes somatosensory temporal discrimination in healthy humans. *PLoS ONE* **7**, e32979 (2012).

13. Chung, S. W. et al. Demonstration of short-term plasticity in the dorsolateral prefrontal cortex with theta burst stimulation: a TMS-EEG study. *Clin. Neurophysiol.* **128**, 1117–1126 (2017).

14. Anderson, R. J., Hoy, K. E., Daskalakis, Z. J. & Fitzgerald, P. B. Repetitive transcranial magnetic stimulation for treatment resistant depression: re-establishing connections. *Clin. Neurophysiol.* **127**, 3394–3405 (2016).

15. Liston, C. et al. Default mode network mechanisms of transcranial magnetic

16. Taylor, S. F. et al. Changes in brain connectivity during a sham-controlled, transcranial magnetic stimulation trial for depression. *J. Affect. Disord.* **232**, 143–151 (2018).

17. Mayberg, H. S. Targeted electrode-based modulation of neural circuits for depression. *J. Clin. Invest.* **119**, 717–725 (2009).

18. Weigand, A. et al. Prospective validation that subgenual connectivity predicts antidepressant efficacy of transcranial magnetic stimulation sites. *Biol. Psychiatry* **84**, 28–37 (2018).

19. Singh, A. et al. Personalized repetitive transcranial magnetic stimulation temporarily alters default mode network in healthy subjects. *Sci. Rep.* **9**, 5631 (2019).

20. Greicius, M. D. et al. Resting-state functional connectivity in major depression: abnormally increased contributions from subgenual cingulate cortex and thalamus. *Biol. Psychiatry* **62**, 429–437 (2007).

21. Li, B. et al. A treatment-resistant default mode subnetwork in major depression. *Biol. Psychiatry* **74**, 48–54 (2013).

22. Manoliu, A. et al. Insular dysfunction within the salience network is associated with severity of symptoms and aberrant inter-network connectivity in major depressive disorder. *Front. Hum. Neurosci.* **7**, 1–17 (2014).

23. Zhu, X. et al. Evidence of a dissociation pattern in resting-state default mode network connectivity in first-episode, treatment-naive major depression patients. *Biol. Psychiatry* **71**, 611–617 (2012).

24. van Tol, M.-J. et al. Local cortical thinning links to resting-state disconnectivity in major depressive disorder. *Psychol. Med.* **44**, 2053–2065 (2014).

25. Sheline, Y. I., Price, J. L., Yan, Z. & Mintun, M. A. Resting-state functional MRI in depression unmasks increased connectivity between networks via the dorsal nexus. *Proc. Natl Acad. Sci.* **107**, 11020–11025 (2010).

26. Alexopoulos, G. S. et al. Functional connectivity in the cognitive control network and the default mode network in late-life depression. *J. Affect. Disord.* **139**, 56–65 (2012).

27. Berman, M. G. et al. Depression, rumination and the default network. *Soc. Cogn. Affect. Neurosci.* **6**, 548–555 (2011).

28. Wu, M. et al. Default-mode network connectivity and white matter burden in late-life depression. *Psychiatry Res. Neuroimaging* **194**, 39–46 (2011).

29. Andreescu, C. et al. Resting state functional connectivity and treatment response in late-life depression. *Psychiatry Res. Neuroimaging* **214**, 313–321 (2013).

30. Bluhm, R. et al. Resting state default-mode network connectivity in early depression using a seed region-of-interest analysis: decreased connectivity with caudate nucleus. *Psychiatry Clin. Neurosci.* **63**, 754–761 (2009).

31. Mulders, P. C. P., van Eijndhoven, P. F., Schene, A. H., Beckmann, C. F. & Tendolkar, I. Resting-state functional connectivity in major depressive disorder: a review. *Neurosci. Biobehav. Rev.* **56**, 330–344 (2015).

32. Wischnewski, M. & Schutter, D. J. L. G. Efficacy and time course of theta burst stimulation in healthy humans. *Brain Stimul.* **8**, 685–692 (2015).

33. Markett, S. et al. Intrinsic connectivity networks and personality: The temperament dimension harm avoidance moderates functional connectivity in the resting brain. *Neuroscience* **240**, 98–105 (2013).

34. Yang, T. T. et al. Adolescent subgenual anterior cingulate activity is related to harm avoidance. *Neuroreport* **20**, 19–23 (2009).

35. Hakamata, Y. et al. Gender difference in relationship between anxiety-related personality traits and cerebral brain glucose metabolism. *Psychiatry Res.—Neuroimaging* **173**, 206–211 (2009).

36. Hakamata, Y., Iwase, M., Kato, T., Senda, K. & Inada, T. The neural correlates of mindful awareness: a possible buffering effect on anxiety-related reduction in subgenual anterior cingulate cortex activity. *PLoS ONE* **8**, e75526 (2013).

37. Hornboll, B. et al. Neuroticism predicts the impact of serotonin challenges on fear processing in subgenual anterior cingulate cortex. *Sci. Rep.* **8**, 1–10 (2018).

38. Kampman, O. et al. Temperament profiles, major depression, and response to treatment with SSRIs in psychiatric outpatients. *Eur. Psychiatry* **27**, 245–249 (2012).

39. Balestri, M. et al. Temperament and character influence on depression treatment outcome. *J. Affect. Disord.* **252**, 464–474 (2019).

40. Rossi, S. et al. Safety, ethical considerations, and application guidelines for the use of transcranial magnetic stimulation in clinical practice and research. *Clin. Neurophysiol.* **120**, 2008–2039 (2009).

41. Lefaucheur, J.-P. P. et al. Evidence-based guidelines on the therapeutic use of repetitive transcranial magnetic stimulation (rTMS). *Clin. Neurophysiol.* **125**, 2150–2206 (2014).

42. Oldfield, R. C. The assessment and analysis of handedness: the Edinburgh

43. Liu, Q., Zhou, R., Chen, S. & Tan, C. Effects of head-down bed rest on the executive functions and emotional response. *PLoS ONE* **7**, e52160 (2012).

44. Moeller, S. et al. Multiband multislice GE-EPI at 7 tesla, with 16-fold acceleration using partial parallel imaging with application to high spatial and temporal whole-brain fMRI. *Magn. Reson. Med.* **63**, 1144–1153 (2010).

45. Setsompop, K. et al. Blipped-controlled aliasing in parallel imaging for simultaneous multislice echo planar imaging with reduced g-factor penalty. *Magn. Reson. Med.* **67**, 1210–1224 (2012).

46. Jenkinson, M., Beckmann, C. F., Behrens, T. E. J., Woolrich, M. W. & Smith, S. M. FSL. *Neuroimage* **62**, 782–790 (2012).

47. Brett, M., Anton, J.-L., Valbregue, R. & Poline, J.-B. in *Neuroimage* Vol. 16, 769–1198 (Academic Press, 2002).

48. Kaiser, R. H., Andrews-Hanna, J. R., Wager, T. D. & Pizzagalli, D. A. Large-scale network dysfunction in major depressive disorder. *JAMA Psychiatry* **02478**, 603–611 (2015).

49. Hansenne, M. & Ansseau, M. Harm avoidance and serotonin. *Biol. Psychol.* **51**, 77–81 (1999).

50. Abrams, K. Y. et al. Trait and state aspects of harm avoidance and its implication for treatment in major depressive disorder, dysthymic disorder, and depressive personality disorder. *Psychiatry Clin. Neurosci.* **58**, 240–248 (2004).

51. Fox, M. D., Buckner, R. L., White, M. P., Greicius, M. D. & Pascual-Leone, A. Efficacy of transcranial magnetic stimulation targets for depression is related to intrinsic functional connectivity with the subgenual cingulate. *Biol. Psychiatry* **72**, 595–603 (2012).

52. Menon, V. Large-scale brain networks and psychopathology: a unifying triple network model. *Trends Cogn. Sci.* **15**, 483–506 (2011).

53. Menon, V. & Uddin, L. Q. Saliency, switching, attention and control: a network model of insula function. *Brain Struct. Funct.* **214**, 655–667 (2010).

54. Fox, M. D. & Raichle, M. E. Spontaneous fluctuations in brain activity observed with functional magnetic resonance imaging. *Nat. Rev. Neurosci.* **8**, 700–711 (2007).

55. Hamilton, J. P. et al. Default-mode and task-positive network activity in major depressive disorder: implications for adaptive and maladaptive rumination. *Biol. Psychiatry* **70**, 327–333 (2011)

56. Bush, G., Luu, P. & Posner, M. I. Cognitive and emotional influences in anterior cingulate cortex. *Trends Cogn. Sci.* **4**, 215–222 (2000).

57. Sridharan, D., Levitin, D. J. & Menon, V. A critical role for the right fronto-insular cortex in switching between central-executive and default-mode networks. *Proc. Natl Acad. Sci.* **105**, 12569–12574 (2008).

58. Iwabuchi, S. J. et al. Alterations in effective connectivity anchored on the insula in major depressive disorder. *Eur. Neuropsychopharmacol.* **24**, 1784–1792 (2014).

59. Cooney, R. E., Joormann, J., Eugène, F., Dennis, E. L. & Gotlib, I. H. Neural correlates of rumination in depression. *Cogn. Affect. Behav. Neurosci.* **10**, 470–478 (2010).

60. Iwabuchi, S. J., Auer, D. P., Lankappa, S. T. & Palaniyappan, L. Baseline effective connectivity predicts response to repetitive transcranial magnetic stimulation in patients with treatment-resistant depression. *Eur. Neuropsychopharmacol.* **29**, 681–690 (2019).

61. Baeken, C., Duprat, R., Wu, G.-R. R., De Raedt, R. & van Heeringen, K. Subgenual anterior cingulate–medial orbitofrontal functional connectivity in medication-resistant major depression: a neurobiological marker for accelerated intermittent theta burst stimulation treatment? *Biol. Psychiatry Cogn. Neurosci. Neuroimaging* **2**, 556–565 (2017).

62. Carter, J. D. et al. Rumination: relationship to depression and personality in a clinical sample. *Personal. Ment. Health* **3**, 275–283 (2009).

63. Manfredi, C. et al. Temperament and parental styles as predictors of ruminative brooding and worry. *Pers. Individ. Dif.* **50**, 186–191 (2011).

64. Lois, G. & Wessa, M. Differential association of default mode network connectivity and rumination in healthy individuals and remitted MDD patients. *Soc. Cogn. Affect. Neurosci.* **11**, 1792–1801 (2016).

65. Huggins, A. A., Belleau, E. L., Miskovich, T. A., Pedersen, W. S. & Larson, C. L. Moderating effects of harm avoidance on resting-state functional connectivity of the anterior insula. *Front. Hum. Neurosci.* **12**, 1–9 (2018).

66. Duprat, R., Wu, G. R., De Raedt, R. & Baeken, C. Accelerated iTBS treatment in depressed patients differentially modulates reward system activity based on anhedonia. *World J. Biol. Psychiatry* **0**, 1–12 (2017).

6 Default mode network and cardiac changes in major depressive disorder after intermittent theta burst stimulation

The results from chapter 4 show that the rsfMRI based personalized left DLPFC target selection is robust and reproducible for clinical use. By applying a single session of 10 Hz rTMS and iTBS in healthy subjects, we reported significant changes in the functional connectivity of the DMN to deeper brain regions like the ventral striatum and the sgACC (10 Hz rTMS) as well as the dACC, rACC, and right AI (iTBS). Furthermore, we found a relationship between DMN connectivity changes and HA scores across both the 10 Hz rTMS and iTBS cohorts, implying a possible role of HA in predicting the rTMS response. While the results certainly elucidate the underlying DMN mechanisms of personalized 10 Hz rTMS and iTBS and their relationship to personality trait of HA, its translation to patients with depression is limited. This is due to aberrant brain connectivity underlying the disease, that act as additional confounds. Several works in the past have explored the feasibility of applying rTMS at prospectively selected personalized left DLPFC sites in patients with depression [247, 248], however none have explored the underlying network mechanisms of such personalized approaches. Hence, we aimed to explore the DMN mechanisms of personalized rTMS in patients with depression. Moreover, to our knowledge, no studies have compared the effects of personalized and non-personalized rTMS either. Considering that such MRI based personalized approaches will put further burden on the health care system (through an additional MRI scan), it is essential to examine the benefits of such personalized rTMS over non-personalized rTMS (e.g. F3 based). Such analysis will further promote a precision medicine approach in the treatment of psychiatric disorders. Therefore, we also aim to compare the effects resulting from our rsfMRI based personalized rTMS approach against the current standard clinical practice of F3 based iTBS.

For treatment of depression, recent work showing iTBS as non-inferior to 10 Hz rTMS [88], led to FDA approval of iTBS for clinical use. The curtailed session duration of iTBS benefits by reducing the time encumberment on patients receiving rTMS treatment. Additionally, patients with depression have higher HA than healthy subjects [184]. As our results from healthy subjects suggest that greater iTBS induced connectivity changes are associated with higher HA scores, we speculated that iTBS may be more beneficial at influencing the underlying brain circuitry than 10 Hz rTMS. Hence, we chose to further explore iTBS mechanisms over 10 Hz rTMS. To further reduce the burden of treatment, we applied iTBS in an accelerated manner (four sessions per day). We expected aiTBS to promote patient compliance for receiving the full treatment. Thus, to explore the mechanism of rsfMRI based personalized aiTBS and compare it against non-personalized aiTBS, we applied aiTBS to two groups of patients with depression using a triple blind (subject, experimenter, and clinical interviewer), crossover, sham controlled study design. One group received the standard F3 based aiTBS while the other received aiTBS at left DLPFC sites chosen based on individual rsfMRI data. Considering the importance of the DMN and its nodes in the pathophysiology of depression [146, 250] and antidepressant response [146, 159–161], we evaluate the DMN. Based on our results of iTBS in healthy subjects, we expected to see a change in functional connectivity of the DMN to the dACC, rACC, and right AI. As sgACC-DMN connectivity is aberrantly higher in patients with depression [146], we hypothesized that aiTBS would additionally reduce sgACC-DMN connectivity post treatment. However, considering the precise nature of this personalized approach, we expected these effects to be more robust in the personalized aiTBS group than in the F3 aiTBS group. Consequently, we also expected higher connectivity changes to translate to higher therapeutic response from personalized aiTBS compared to F3. Finally, based on the relationship observed in healthy subjects, we expected that the changes induced by aiTBS in patients with depression will have a positive correlation to HA scores. Note, another novel method for personalizing rTMS at the left DLPFC is the NCG-TMS method. It advocates applying rTMS to those left DLPFC sites that elicit maximum HR changes. However, the prospective utility of NCG-TMS in the treatment of depression remains to be ascertained. As our method prospectively selects optimal left DLPFC sites based on its connectivity to the sgACC, we hypothesise that iTBS at such sites would also elicit larger HR changes as compared to the relatively imprecise F3 approach. Hence, we acquired ECG data to investigate the HRV and HR effects of personalized and F3 iTBS approaches. We hypothesize that the HRV and HR influences resulting from personalized approaches would be stronger than those resulting from the F3 approach. Further details on the methodology used, the outcomes of the experiment and a corresponding discussion of the results are presented below.

My contributions pertain to formulating the study design, data collection and analysis and writing the final draft of the manuscript presented.

Default mode network and cardiac changes in major depressive disorder after intermittent theta burst stimulation

*Aditya Singh[1], Tracy Erwin-Grabner[1], Walter Paulus[2], Niels Hansen[1], Carsten Schmidt-Samoa[3], Peter Dechent[3], Andrea Antal[2, 4], Roberto Goya-Maldonado[1,] **

Affiliations:

[1]Systems Neuroscience and Imaging in Psychiatry, Department of Psychiatry and Psychotherapy of the University Medical Center Göttingen

[2]Department of Clinical Neurophysiology of the University Medical Center Göttingen

[3]Department of Cognitive Neurology of the University Medical Center Göttingen

[4]Department of Neurology of the University Medical Center Göttingen

*To whom correspondence should be addressed:

Dr. Roberto Goya-Maldonado

Laboratory of Systems Neuroscience and Imaging in Psychiatry (SNIP-Lab)

Department of Psychiatry and Psychotherapy

University Medical Center Göttingen

Von Siebold Straße 5, 37075, Göttingen, Germany

Tel: +49 (0) 551-39-22244

Email: roberto.goya@med.uni-goettingen.de

Introduction

High frequency repetitive transcranial magnetic stimulation (rTMS) delivered at the left dorsolateral prefrontal cortex (DLPFC) has become an established therapy option for the treatment of depression (1). Intermittent theta bust stimulation (iTBS) is a newer protocol, increasingly used in the clinical practice (2–6) due to shorter duration and comparable efficacy to the first FDA-approved 10 Hz rTMS protocol. Nevertheless, a considerable portion of depressed patients still do not benefit from rTMS, probably because of anatomical and functional inter-individual variabilities in the selection of the optimal site of stimulation, which have not yet been prospectively integrated to treatment protocols. Retrospective studies have shown that success of treatment may indeed be related to the optimal site of stimulation from patient to patient (7–9). Based on this evidence we had established and presented a method of personalizing rTMS by selecting individual sites to deliver stimulation based on the participant's resting state fMRI (rsfMRI) data with a high precision of coil localization (10). We then validated this target selection method and investigated the effects of iTBS on relevant brain networks in healthy individuals (11). To truly usher the era of systems medicine, one must empirically contrast the benefits of such personalized treatments (e.g. the rsfMRI based personalization) over the available standard approach (e.g. the Beam F3 method). Arguably, the delivery of iTBS at pertinent network nodes will aid better propagation of iTBS effects to further distal regions in the brain. Hence, we expected such effects to translate to greater therapeutic response following personalized iTBS in relation to using standard protocols targeting F3. To address the functional variabilities, some previous works have presented evidence on the feasibility of other rsfMRI based personalized approaches in treatment of patients with depression (12,13). However, these studies have not explored the underlying mechanism of such personalized rTMS. Therefore, we intended to investigate here the mechanisms of the personalized iTBS in patients with depression and to systematically compare this approach to the standard F3 (10-20 EEG system) iTBS. The F3 approach was developed to more reliably target the left DLPFC in comparison to the 5 cm rule, which is stimulating ~5.5 cm anterior to motor-cortex (14). Though better than the 5 cm rule, the F3 method still fails to address differences in the functional architecture of the DLPFC, much like the left DLPFC MNI coordinate based stimulation (10).

The default mode network (DMN) consists of the posterior cingulate cortex (PCC), medial prefrontal cortex, and parts of posterior parietal cortex (15), which seems consistently hyperconnected to the subgenual anterior cingulate cortex (sgACC) in depression (15,16). A reduction of this hyperconnectivity has been related to a reduction of depressive symptoms (17). Using 10 Hz rTMS, a study showed that aberrant sgACC connectivity with the DMN decreases with treatment in responders but not in non-responders (15). Given the central involvement of the DMN in the pathophysiology of depression and response to treatment, our current investigation aims to explore the underlying mechanism of personalized iTBS on the DMN in patients with depression. A single session of iTBS (1800 pulses) in healthy subjects using the rsfMRI based personalization method reduces the connectivity of the DMN to the rostral anterior cingulate cortex (rACC), dorsal anterior cingulate cortex (dACC), and right anterior insula (rAI) (11). Therefore, we hypothesize that personalized stimulation over left DLPFC targets will influence the FC of DMN with rACC, dACC, and rAI in patients with depression. Considering probable elevated sgACC-DMN connectivity in patients, we expect a reduction in sgACC-DMN connectivity as well. Using an accelerated iTBS (aiTBS) protocol, Baeken et al. (18) found that the FC of the sgACC changed in iTBS responders. In addition, they found that the iTBS induced increase in FC of sgACC to medial orbitofrontal cortex (mOFC) correlated with a decrease in hopelessness, suggesting a relationship between behavioural domains and FC changes driven by iTBS. In the same sample of patients, Duprat et al. (19) reported distinctive effects of iTBS on the neural activity reward system based on pre-treatment anhedonia ratings. The role of behavioural and psychological factors in predicting rTMS response has been reported in past work (20). The ease of obtaining such behavioural measures makes them preferable tools to predict treatment response. Pertaining to antidepressant response, a recent meta-analysis showed that personality dimensions are associated with depression treatment outcomes (21,22). Several studies have also shown the predictive associations of personality and antidepressant response to rTMS (23–25). We also reported a relationship between the harm avoidance (HA) trait from the temperament and character inventory (TCI) and FC changes induced by single sessions of both 10 Hz rTMS (10) and iTBS (11) in healthy adults. Therefore, we theorize that HA would be associated with FC changes and therapeutic response after application of aiTBS in patients with depression.

In addition to the neuroimaging information, we acquired online electrocardiogram (ECG) data during iTBS and during rsfMRI in depressed subjects. We calculated heart rate variability (HRV) from ECG data, which is a measure of beat to beat intervals (26). It estimates the functioning of the autonomic nervous system (ANS)(27) and is a marker for flexible dynamic regulation of autonomic activity (28). The ANS is composed of the parasympathetic and sympathetic system. It is usually under the influence of a several brain regions that together form what is referred to as the central autonomic network (CAN)(29). The CAN involves brain regions like the hypothalamus, insula, anterior cingulate cortex, and amygdala (30). These regions allow integration of emotional responses with sensations in the body (29). An adequate HRV allows parasympathetic control over emotions and their regulation in individuals (30) and can be viewed as a dynamic balance between the parasympathetic and sympathetic system. A meta-analysis of PET and fMRI studies relating HRV and neural structures showed that regions like the perigenual ACC (pgACC), sgACC and the ventromedial prefrontal cortex (vmPFC) are associated with HRV (28). These regions, as we know, are typically involved in regulation of emotions (28,31–33). It is also worth noting that depression increases the risk of cardiovascular malfunction approx. 2-5 folds, indicating an interaction between depressive disorder and the heart-brain axis (27). Studies have shown that patients with depression show lower HRV and a higher heart rate (HR), when compared to healthy individuals (34,35), and that neuromodulation techniques like rTMS can increase HRV and decrease HR in patients with depression (36). Recently, a study applied a 3-minute iTBS protocol in patients with depression and reported significant increase in HRV and reduction in HR in real iTBS condition but not in sham iTBS (37). These influences could be observed within the first 30 seconds of iTBS, pointing towards the possibility of using immediate cardiac changes as a marker for appropriate target engagement of the sgACC via left DLPFC. As we are using rsfMRI to locate personalized targets in left DLPFC for stimulation that most likely interact with the sgACC, we hypothesize that our method would induce greater HRV and HR changes than the standard F3.

Here we applied four daily sessions of iTBS at left DLPFC over five weekdays. We analyzed the DMN before and after the weeklong stimulation in a triple-blind (patients, clinicians, experimenter), sham-controlled, and crossover (real and sham in pseudo-randomized order) study, similar to (3). To our knowledge such personalization based on individual large-scale networks (10,11) at left DLPFC have not been previously reported in patients with depression. In addition, no study has investigated the F3 and personalized iTBS effects on the DMN, extending to HRV effects, in the same sample. We expect that depicting the effects in patients with depression will more concretely highlight the benefits of personalized approaches. If personalized iTBS leads to stronger symptomatic reduction, stronger FC modulation of the sgACC, and greater HRV increase than F3 iTBS, we hope it will consolidate a first step towards precision medicine in psychiatry.

Materials and Methods

Participants

We enrolled patients aged 18-60 years, diagnosed with major depressive disorder (MDD) or bipolar disorder (BD), and currently presenting moderate to severe symtpoms. Trained clinicians evaluated the participants with structured clinical interviews (see below). The study is in agreement with relevant guidelines and regulations (1,38). The Ethics Committee at the University of Medical Center Göttingen approved the study protocol and participants provided informed consent prior to participation.

Study design

The study is a sham-controlled, triple-blind (subject, experimenter, and clinical interviewer), crossover study with daily aiTBS (4 sessions/day) delivered in a pseudo-randomized and counterbalanced fashion. The investigation was completed in 15 visits spread over a span of six weeks, with weekly evaluations of pre- and post-intervention (Fig 1).

Visit 0 (Week 0)

On visit 0, a medical doctor determined the suitability of the participant for the study based on inclusion and exclusion criteria. Inclusion criteria were diagnosis based on the structured clinical interview for DSM − 5 (SCID-5) and the actual severity of depressive symptoms (only moderate or severely depressed patients were included). Past information on presence of other comorbid disorders that could mimic or hinder a clear assessment of depressive symptomatology was considered exclusion criterion, e.g. borderline personality disorder, presence of significant psychiatric illness comorbid with depression (except for anxiety), contraindication to study protocols (rTMS, fMRI) such as metal implants (except dental fixtures), individual and family history of epilepsy or unexplained seizures, presence of unprescribed drugs (e.g. cannabis) or recent change in medication regime, if taking psychotropic medications. The clinician explained the study protocol to the participants after which they gave their informed consent. As part of a screening to exclude other general causes of depressive symptoms (e.g. hypothyroidism, cocaine), the clinician then drew their blood sample and participants deposited their urine sample. The blood sample also confirmed circulating levels of regular medication intake, if any. Upon inclusion, patients were instructed to continue their current medications without change in type or dose during the course of the study (i.e. until the end of week 6), unless medically indicated. Subjects were randomly allocated to receive either standard F3 or personalized stimulation.

Visit 1 (Week 1)

On visit 1, the patients completed the Beck Depression Inventory II (BDI II), Barratt Impulsiveness Scale (BIS), Beck Anxiety Inventory (BAI), TCI questionnaire, the Childhood trauma questionnaire (CTQ), and the World Health Organization Quality of Life Scale (WHOQOL-Bref). They further filled out a handedness questionnaire (39), and a vocabulary-based intelligence test (MWTB)(40,41) in addition to a "daily sheet" questionnaire, detailing their coffee and nicotine consumption over the past two hours, alcohol consumption and suicide ideation over the past 24 hours, and their sleep satisfaction from the previous night. The patients met with trained clinicians blind to the interventions, who administered the Montgomery-Asberg Depression Rating Scale (MADRS), Hamilton Depression Rating Scale (HAM-D) and Young Mania Rating Scale (YMRS). Clinicians also collected the blood sample to monitor medication intake. After this, the experimenters explained the MRI procedure to the patients and familiarized them with the MR scanner session. We acquired T1 weighted and rsfMRI data (first day, R1). When the subjects were allocated to personalized treatment, the R1 rsfMRI data was used for calculation of personalized left DLPFC targets. This target was utilized for stimulation during visit 2 through visit 6.

Visit 2 – 6 (Week 2)

Subjects visited the lab for daily stimulation sessions (5 days). Subjects were seated in a reclined, comfortable chair, where we first measured their resting motor threshold (RMT) to determine the minimum intensity of the stimulator output at which five out ten times an electromyograph (EMG) recording of more than 50 µV was obtained. Based on this, the stimulation intensity for the daily sessions was set at 110% of the RMT. After this daily measurement, we began the iTBS stimulation session which consisted of four iTBS sessions with twenty-minute pause intervals in between (18). During each iTBS session, we additionally acquired ECG data using the neuroConn NCG-rTMS device (see ECG acquisition for details). Before and after daily stimulation, the subjects filled out the Positive and Negative Affect Schedule (PANAS) questionnaire, and the *daily sheet*, including a survey inquiring about any potential discomfort experienced from stimulation sessions. After the last daily session of stimulation, subjects underwent their second MRI scan. Comparable to the first visit (visit 1), we again

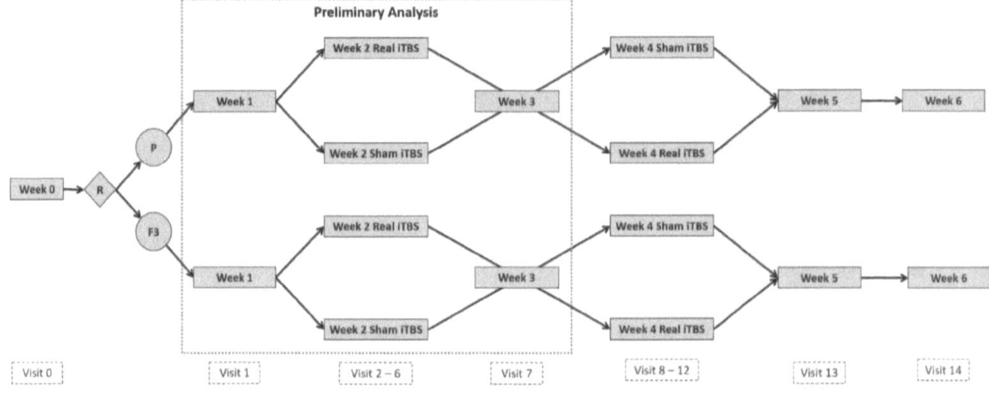

Figure 1: The brief schematic explaining the study design. Upon determining inclusion/exclusion criteria during week 0 (visit 0), the participants are randomly allocated to receive either personalized or F3 iTBS. We acquired rsfMRI data during week 1 (at visit 1), week 2 (at visit 6), week 3 (at visit 7), and week 4 (at visit 12). If slated for personalized iTBS, the rsfMRI data from week 1 was used for target localization for week 2 stimulation and rsfMRI data from week 3 was used for target localization for week 4 stimulation; using method detailed in (10). We obtained depression rating scales (clinician administered: HAM-D, MADRS; self-reported: BDI-II) from week 1 through week 6 (at visit 1, visit 6, visit 7, visit 12, visit 13, and visit 14). The rsfMRI data collected within the red box (week 1 till week 3) have been included in preliminary analysis. R: randomization, P: personalized stimulation, F3: standard stimulation, iTBS: intermittent theta burst stimulation

acquired T1 and rsfMRI images (second day, R2). After this the subjects met with the clinician, who again administered the HAM-D, MADRS, and YMRS scales. The subjects completed the BDI-II, BAI, and BIS scale in reference to symptoms experienced during the week prior. An additional blood sample was obtained, marking the end of the first stimulation week.

Visit 7 (Week 3)

After a weeklong pause from stimulation, patients visited the lab for visit 7, which included completing the self-report measures BDI-II, BAI, BIS and *daily sheet* in addition to meeting with the clinician, who administered the HAMD, MADRS and YMRS scales. Blood samples were again taken. The rsfMRI data (third day, R3) was used for calculation of personalized left DLPFC target for the following week (visit 8 through 12) when the subject was allocated to personalized treatment.

Visit 8 – 12 (Week 4)

Visits 8 – 12 were identical to visits 2 – 6, except for the type of stimulation (real/sham).

Visit 13 (Week 5)

Visit 13 was approximately one week after visit 12. The subjects completed the BDI-II, BAI, BIS, and the *daily sheet*. The clinician administered the HAM-D, MADRS, YMRS and the Columbia Suicide Severity Rating Scale. Lastly, a final blood sample was obtained.

Visit 14 (Week 6)

Visit 14 again included clinician administered HAMD, MADRS and YMRS, as well as the *daily sheet,* the BDI-II, BAI and BIS self-report measures and the addition of the WHOQOL-Bref.

Intermittent TBS Protocol

We delivered iTBS using a MagVenture X100 in research study mode with a "figure of 8" MCF-B65 A/P cooled coil. As part of the stimulation setup, two surface electrodes were attached to the patient's head, as close to the site of stimulation as possible. These electrodes deliver small current stimulation, mimicking the skin and muscle sensation experienced during real rTMS. They were applied during both the real and sham conditions, so that the setup was identical for real and sham rTMS and hence indistinguishable for patients and experimenters. We used stimulation parameters for a 1800 pulses iTBS session as described elsewhere (5). In short, it consisted of 3 pulses bursting at 50 Hz at the frequency of 5 Hz for 2 seconds with an 8-second intertrain interval, totalling 60 trains delivered for 9 minutes 40 seconds. The real or sham application of the coil was determined by a six-digit code that dictated which coil orientation must be used for stimulation. The code and the corresponding coil orientation (real stimulation or sham stimulation) was only available to the study leader (RGM), who was not directly involved in any steps of intervention or clinical rating. The subjects were pseudo-randomly allotted to receive either real first (week 2) and sham later (week 4) or sham first (week 2) and real later (week 4). On each visit during week 2 and week 4 (visits 2-6 and visits 8-12, respectively), the subjects received four sessions of 1800 pulses iTBS over the left DLPFC (F3 or personalized). As previous works show that therapeutic gains of multiple session rTMS is related to number of sessions of rTMS rather than the total number of pulses (42), we applied the four sessions of iTBS separated by twenty-minute pauses, the length of which was adapted from Baeken et al. (18). The subjects were given ambulatory freedom during the pauses but instructed not to consume nicotine or caffeine at that time nor for two hours prior the start of stimulation.

Image Acquisition

We collected structural data (T1-weighted scans with 1x1x1 mm resolution) with a 3T MR scanner (MAGNETOM Prisma, Siemens Healthcare, Erlangen, Germany) using a 64-channel head coil. We acquired the rsfMRI data using the T2*-weighted gradient-echo echo-planar imaging sequence provided by the Center for Magnetic Resonance Research of the University of Minnesota (43,44). It had the following parameters: repetition time of 1.5 s, echo time of 30 ms, flip angle of 70°, 69 axial slices with a multi-band factor of 3, 2x2x2 mm, FOV of 189 mm, with 10% gap between slices and posterior to anterior phase encoding. The rsfMRI data were acquired with 400 volumes in 10 minutes. The gradient-echo field map was acquired with a repetition time of 704 ms, echo times of 4.92 ms (TE 1) and 7.38 ms (TE 2), flip angle of 60°, 73 slices, FOV of 210 mm, 2x2x2 mm, with 10% gap between slices and anterior to posterior phase encoding. Using the physiological monitoring unit, we simultaneously recorded ECG (400 Hz), respiration (50 Hz), and pulse (50 Hz) data.

Imaging Data Analysis

We preprocessed the individual rsfMRI data using SPM12 and MATLAB 2015b (The MathWorks, Inc., Natick, MA, USA) to execute the following state-of-the-art steps: slice time correction, motion correction, gradient echo field map unwarping, normalization, and regression of motion nuisance parameters, cerebrospinal fluid, white matter, and physiological noise (respiration: breathing belt and heart rate: ECG). Following this, we temporally concatenated the data for group independent component analysis (ICA) with FSL 5.0.7 software (45). We visually identified the independent component (IC) that best resembled the DMN. We back reconstructed this IC representing the DMN in the normalized rsfMRI data of individual subjects and r-to-z transformed for statistical analysis.

ECG Acquisition

We acquired ECG data during online iTBS using the neuroConn NCG-rTMS device (neuroCare, Munich, Germany). The device provides an easy to use interface to record ongoing ECG activity by placing three electrodes to the breast region of the participants. The ground electrode (black) is placed close to the lower end of the sternum; the red electrode is placed in the 1st intercostal space on the right-hand side. The green electrode is placed in the 4th or 5th intercostal space on the left side. The device displays the current heart rate and stores the raw ECG data into an SD card in European Data Format+ (EDF+). The device records the ECG signal at 1000 Hz as well as the exact triggers from the rTMS device, thus facilitating synchronization of ECG and iTBS bursts.

ECG Data Analysis

We analysed the ECG data using the Biosignal Processing in Python (BioSPPy) toolbox, the HRV analysis toolbox, and the SciPy module (46) in Python 3.6. We used SciPy to remove powerline noise from the ECG signal, then used BioSPPy to remove band wander noise in the data and detect R-peaks. We further corrected the automatically detected R-peaks to match with the maxima of the curve with a tolerance of 50 milliseconds. After this, we calculated the RR interval using the time stamps of detected R-peaks. We then preprocessed the RR interval by developing a Python implementation of the algorithm detailed in Lipponen and Tarvainen, 2019 (47). This implementation allows for correction of ectopic beats and other types of outliers. Following this, the RR interval data was split into various segments of varying lengths (30 s, 45 s, 60 s, 180 s, 270 s, 360 s and the full length of the stimulation duration). Normally, the deceleration of HR is visible rather immediately following iTBS. For this reason, the slope of the RR interval plotted against time was calculated for 30 s, 45 s, and 60 s. A positive slope would indicate increased RR intervals, implying a decrease in HR during the time frame. We then calculated the HRV parameters using the HRV analysis toolbox mentioned above.

The HRV can be quantified using either time domain or frequency domain measures. Here we have presented results using the time domain measures, which provide a higher consensus regarding their interpretability. Thus, we present results only from time domain measures, as the frequency domain did not offer additional advantages. For example, there is still some debate regarding the influence of parasympathetic and sympathetic systems on the low frequency component of HRV (LF-HRV). Conversely, the high frequency component of HRV (HF-HRV) is rather consistent to quantify parasympathetic influence on the ANS, but this parameter correlates well with the time domain measure of RMSSD (48). Therefore, we extracted the mean RR interval, root mean square of successive differences (RMSSD), and the standard deviation of the NN (RR) intervals (SDNN). The RMSSD and SDNN are easily calculated from the RR interval information. The RMSSD is indicative of the parasympathetic influence of the ANS and reduced values of RMSSD suggest reduced parasympathetic activity (49). The SDNN indexes the total variability of the heart beats and is influenced by both the parasympathetic and sympathetic system (49). We applied the natural logarithm transformation to normalize the RMSSD and SDNN values as done previously (37).

Statistical Analysis

a) Depression Rating Scales and PANAS scores

We analysed the depression rating scales (HAM-D, MADRS, and BDI-II) using the SciPy module (46) of Python 3.6. We denoted subjects with a decrease of 50% or more in depression severity scores as responders. The percentage of responders in both F3 and personalized groups was also calculated. In addition, the correlation between BDI-II scores and HRV indices was evaluated using Python SciPy module. All results are evaluated with Bonferroni corrected thresholds. For comparison of symptom scores from week 1 to week 6, a total of six pairwise t-tests are conducted for BDI-II, HAMD, and MADRS (two each for personalized and F3 groups), leading to $p_{Bonferroni}$ = 0.0083. For comparing the changes in symptom scores between personalized and F3 groups, a total of three two-sample t-tests are conducted, resulting in $p_{Bonferroni}$ = 0.017. PANAS scores are also analysed using the SciPy module in Python 3.6. We conduct eight one-sample t-tests to evaluate if the changes in positive and negative affects scores is different from null (real and sham (x2), across positive and negative (x2) affect from personalized and F3 groups (x2)) leading to $p_{Bonferroni}$ = 0.00625. Eight additional two sample t-tests are computed to compare the changes in positive and negative affect between personalized and F3 groups ($p_{Bonferroni}$ = 0.00625). Results surviving Bonferroni correction for multiple comparison are accordingly highlighted **.

b) Imaging

We used the first three rsfMRI scans (R1 [before week 2 iTBS], R2 [after week 2 iTBS], and R3 [during week 3, before week 4 iTBS]) to evaluate the effects of iTBS. We deliberately excluded R4 rsfMRI (after week 4 iTBS), as this dataset would potentially have carryover effects from the previous stimulation week, most likely arising from real iTBS. Considering the modest sample sizes when stratifying subgroups, we refrained from including R4 rsfMRI, which could induce noise in the R1-R4 comparisons due to variable effects on R4. Hence, we compared the R1 (week 1), R2 (week 2), and R3 (week 3) DMN and have excluded R4 (week 4) DMN in this preliminary analysis. In this manner, the interpretation can be simplified to consider each group as receiving either real or sham iTBS and being followed post iTBS for immediate effects (R2) and relatively longer effects (R3). As these two groups do not cross-over, we necessarily included sex, age and the BDI – II scores as covariates-of-no-interest in our ANOVAs. To identify the effects of F3 and personalized iTBS in the DMN, we analysed both groups using individual ANOVAs in SPM12. Using the factorial design ANOVA, we compared the rsfMRI across real and sham conditions. The ANOVAs had two factors, first factor corresponded to the order of stimulation i.e. real iTBS first or sham iTBS first. The second factor determined the rsfMRI (R1, R2, R3) data. Note that R1, R2, R3 are corresponding to week 1 (visit 1), week 2 (visit 6), and week 3 (visit 7) respectively. Figure 2 schematically depicts the analysis model as created in SPM12. We further evaluated the relationship between depression severity scores and the DMN by running a simple linear regression model in SPM12. The results presented are whole brain uncorrected, p < 0.001, and k > 13.

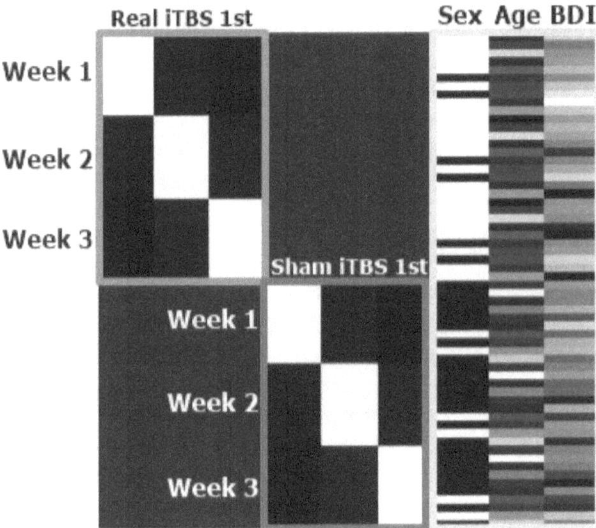

Figure 2: The statistical model of analysis in SPM12. In green are defined week 1, 2, and 3 of those that received real iTBS and in red, week 1, 2, and 3 of those that received sham iTBS. Since, this preliminary analysis does not extend into the crossover step (week 4), sex, age and the respective BDI-II scores were used as covariates-of-no-interest (in yellow).

c) ECG

The HRV parameters were exported to CSV format and further analysed using SPSS version 25 (IBM Corp., Armonk, N.Y., USA). As in the case of imaging data, we analysed the personalized and F3 iTBS group using distinct ANOVAs. We conducted multivariate ANOVAs, corresponding to each time duration (30 s, 45 s, 60 s, 180 s, 270 s, 360 s, and full length): slope; mean RR interval, natural logarithm of RMSSD and natural logarithm of SDNN. For comparison of slope, a total of six tests are computed (personalized and F3 group across 30s, 45s, and 60s) resulting in $p_{Bonferroni}$ = 0.0083. For comparison of mean RR interval, a total of eight tests are computed (personalized and F3 group across 180s, 270s, and 360s, full length) resulting in $p_{Bonferroni}$ = 0.00625. For comparison of RMSSD and SDNN, a total of ten tests are computed (personalized and F3 group across 60s, 180s, 270s, and 360s, full length) resulting in $p_{Bonferroni}$ = 0.005. Results surviving Bonferroni correction for multiple comparison are accordingly highlighted **.

Extraction of sgACC parameter estimates (functional connectivity strengths)

To evaluate the relationship between sgACC-DMN connectivity and HRV indices, we extracted the sgACC parameter estimates using a 5 mm diameter spherical ROI based on independent coordinates from Tik et al. (50).

Relationship between HRV and BDI-II scores

We correlate the HRV parameters (RMSSD and SDNN) with BDI-II scores using the SciPy module (46) in Python 3.6. Two correlation tests are computed for each HRV parameter (personalized and F3) resulting in $p_{Bonferroni}$ = 0.025.

HA as a predictor of iTBS response

A correlation test between HA and symptom changes as well as between sgACC-DMN FC changes is calculated using the SciPy module (46) in Python 3.6. For each modality, symptom change and sgACC-DMN changes, three correlation tests are computed for each of the symptom scales (BDI-II, HAMD, MADRS) resulting in $p_{Bonferroni}$ = 0.017.

Visual analogue scales

At the end of the second stimulation week, we asked all subjects to record the intensity of physical sensations arising from each week of stimulation (week 2 and week 4) by marking a cross on a 10 cm line segment (the left most point implying a score of zero (no sensations or expectations) and the right most point implying a score of ten (maximum sensations or expectations). The same was done to record their expectations of effects on mental states in response to each stimulation type. We compared the reported scalp sensation and expected mental effects from real and sham iTBS using pairwise t-tests ($p_{Bonferroni}$ = 0.025).

Data availability statement

Owing to restrictions in the data sharing consent obtained from the participants of the study, the datasets generated and analysed cannot be made publicly available.

Results

Demographics

Only participants with MDD were included in the preliminary analysis. Of the 62 participants screened for inclusion, 17 subjects dropped out before beginning stimulation mostly due to their inability to commit six-week time for the study as a reason for discontinuing. Forty-five subjects began with aiTBS of which seven subjects dropped out after the start of stimulation (2 from personalized group and 5 from F3 group). Of these seven subjects, two subjects reported migraine and headaches as their respective reason for discontinuing the study (both belonging to F3 group). The dropout rates amongst subjects who began with aiTBS are 9.1% (personalized) and 21.74% (F3). A total of 24 subjects dropped out of the study with 37.5% reporting time as their reason. Thus, 38 subjects finished the study (20 in personalized and 18 in F3 iTBS). The mean age of the 38 participants is 35.8 ± 13.7 years (16 females). Of the 38 participants, 20 received personalized iTBS (real iTBS first, n = 10, sham iTBS first, n = 10) and 18 received F3 iTBS (real iTBS first, n = 10, sham iTBS first, n = 8). The mean depression severity scores at week 1 were as follows: BDI-II 32.47 ± 9.73 (Personalized iTBS group: 30.25 ± 10.77, F3 iTBS group: 34.94 ± 8.02, t = -1.5, p-value = 0.13); HAM-D 15.21 ± 5.29 (Personalized iTBS group: 16.4 ± 5.9, F3 iTBS group: 13.89 ± 4.31, t = 1.5, p-value = 0.14); MADRS 23.58 ± 7.85 (Personalized iTBS group: 23.6 ± 9.4, F3 iTBS group: 23.55 ± 6.03, t = 0.017, p-value = 0.98).

Depression Rating Scales

BDI – II

Comparison of BDI-II scores from the beginning to the end of the study indicates that both the personalized iTBS and F3 iTBS subject groups showed a significant decrease in symptoms (Figure 3; for both personalized (t = 5.47, p-value < 0.001**) and F3 iTBS groups (t = 6.39, p-value < 0.001**, two sided pairwise t-tests). Over the course of six weeks, a trend towards higher total BDI – II score

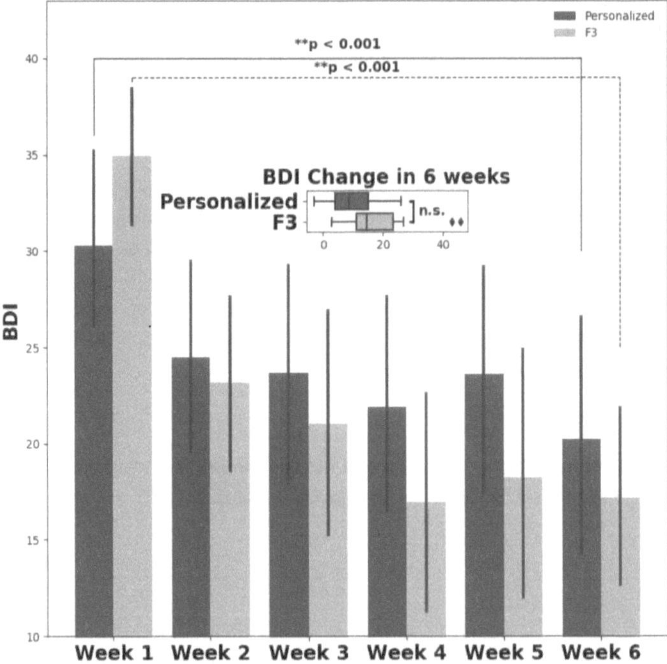

Figure 3*: Weekly BDI-II scores for both personalized and F3 iTBS group. In both groups, there is a significant decreased BDI-II score in week 6 compared to week 1. The total decrease observed during this period is slightly higher in F3 group, although this does not survive correction for multiple comparisons.*

decrease in F3 group is observed (17.78 ± 11.8) compared to personalized group (10.05 ± 8.21) (t = -2.32, p-value = 0.0217, two-sided t-test). This test does not survive multiple comparison corrections. To evaluate the therapeutic response, we measured the proportion of patients showing more than 50% decrease in the BDI-II scores. The amount of treatment responses is so far comparable between personalized and F3 iTBS group, 40% and 38.9%, respectively.

HAM-D

The HAM-D ratings were significantly lower in week 6 compared to week 1 for the personalized and show a trend of decrease in the F3 iTBS groups (two-sided pairwise t-tests, t = 3.63, p-value = 0.0017** and t = 2.82, p-value = 0.0117, respectively). The average change in HAM-D ratings were not significantly different between the two groups (t = 0.60, p-value = 0.55, two-sided t-test). The therapeutic response (more than 50% decrease on HAM-D) was comparable across both groups (personalized: 40%, F3: 33.3%).

MADRS

The MADRS scores also decreased significantly for both groups by week 6 when compared against week 1 (personalized: t= 3.43, p-value = 0.0028**, and F3 iTBS groups: t = 3.5, p-value = 0.0027**, two sided pairwise t-tests). The mean change reported over the course of six weeks was not significantly different between personalized and F3 iTBS groups (t = 0.64, p-value = 0.53, two-sided t-test). See supplementary figure 1 for an overview of HAM-D and MADRS ratings over the study duration. The

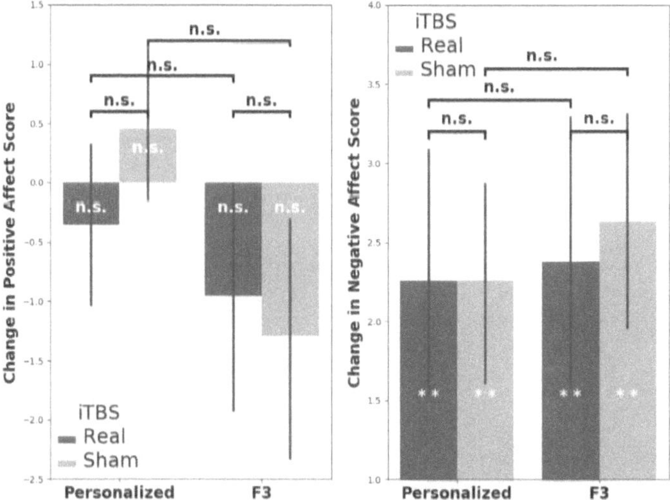

***Figure 4**: Changes in positive and negative PANAS affect scores observed each day of stimulation. The negative affect score changes significantly in both groups (personalized and F3). However, no significant changes in positive affect scores are observed across intervention groups, i.e. there are no significant differences in the changes induced in negative affect by either real or sham iTBS, neither are the changes induced by personalized or F3 iTBS significantly different from each other. [**p-value < 0.05]*

therapeutic response (more than 50% decrease on MADRS) was comparable across both groups (personalized: 30%, F3: 33.3%).

PANAS

We acquired PANAS scores from participants, both before and after daily iTBS sessions. The goal was to track subtle changes in the mood of participants over the course of multiple iTBS sessions each day. We divided the overall PANAS scores into two parts: a positive affect score and a negative affect score. We then compared the changes (Pre[iTBS] – Post[iTBS]) in positive and negative affect scores corresponding to each day of stimulation. We report no significant changes in positive affect scores either from real or sham iTBS (personalized real iTBS: t = -1.037, p-value = 0.3019; personalized sham iTBS: t = 1.36, p-value = 0.176; F3 real iTBS: t = -1.931, p-value = 0.056; F3 sham iTBS: t = -2.41, p-value = 0.018). However, the negative affect score reduced significantly post stimulation. We report these effects across both personalized (real iTBS: t = 5.72, p-value < 0.05**; sham iTBS: t = 6.78, p-value < 0.05**) and F3 (real iTBS: t = 5.92, p-value < 0.05**; sham iTBS: t = 7.41, p-value < 0.05**; two-sided, one-sample t-tests) groups. As evident, the decrease in negative affect is observed across both real and sham iTBS conditions (personalized iTBS, real versus sham: t = 0.005, p-value = 0.99; F3 iTBS, real versus sham: t = -0.47, p-value = 0.635; two-sided pairwise test). There are no differences in the negative affect reduction caused by real personalized iTBS or real F3 iTBS (t = -0.211, p-value = 0.833) or those caused by sham personalized iTBS or sham F3 iTBS (t = -0.77, p-value = 0.44) (Figure 4).

Imaging

Personalized iTBS

A comparison of the DMN across week 1 (R1) and week 3 (R3), correcting for sham iTBS effects, shows that parts of sgACC, PCC, and pre/cuneus display reduced FC to the DMN (p-value < 0.001, k > 13, whole brain, uncorrected) (Figure 5a and b). We do not see these reductions in FC during R1-R2 comparison. Furthermore, R2-R3 comparison also shows reduced functional connectivity within the same regions, albeit to a lesser extent within the PCC. (Figure 5c).

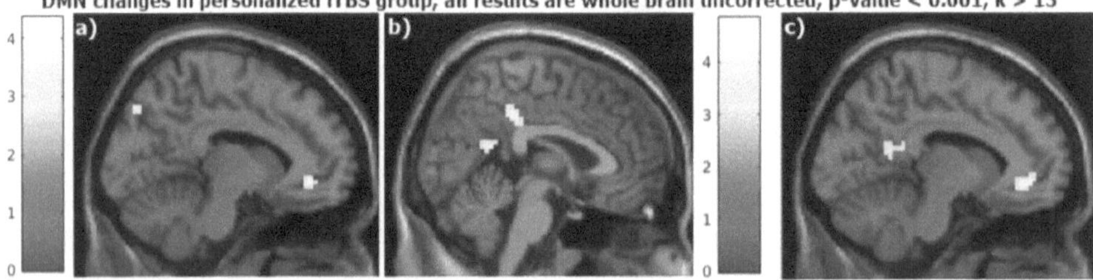

Figure 5: DMN associated with personalized iTBS. a) & b) A comparison of R1 versus R3 indicates a reduced FC between DMN and sgACC, PCC, and precuneus. c) A comparison of R2 versus R3 revealed that the FC reductions observed in R1-R3 comparison can also be seen in R2-R3 comparison. All comparisons showed are corrected for sham effects. Colorbars represents the t-value.

F3 iTBS

Unlike the personalized iTBS group, R1-R3 comparison shows a reduced FC between the DMN and precuneus region but not in the PCC or the sgACC (Figure 6a). However, when comparing the R2-R3 DMN, the sgACC also shows a reduced FC to the DMN (Figure 6: b, c), like the observations from our personalized iTBS group. See supplementary table 1 for details on the coordinates and cluster size of the reported results.

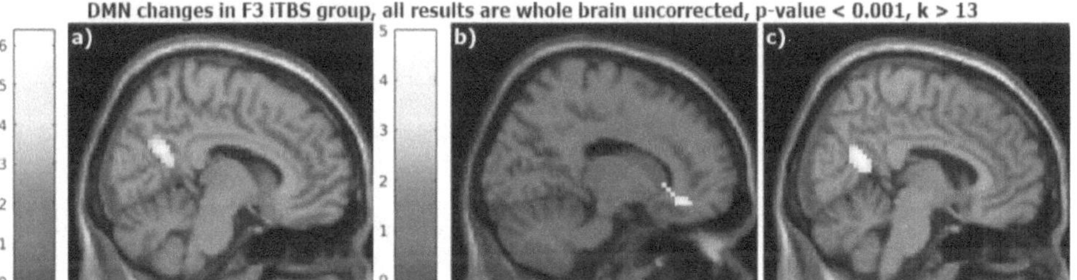

Figure 6: DMN associated with F3 iTBS. a) An R1-R3 comparison in F3 group reveals that parts of precuneus as the only regions showing reduced FC to DMN. Reduction in sgACC FC is not observed in the R1-R3 comparison. b) & c) sgACC and precuneus both show a reduced FC between R2-R3 in F3 iTBS group. All comparisons showed are corrected for sham effects. Colorbars represents the t-value.

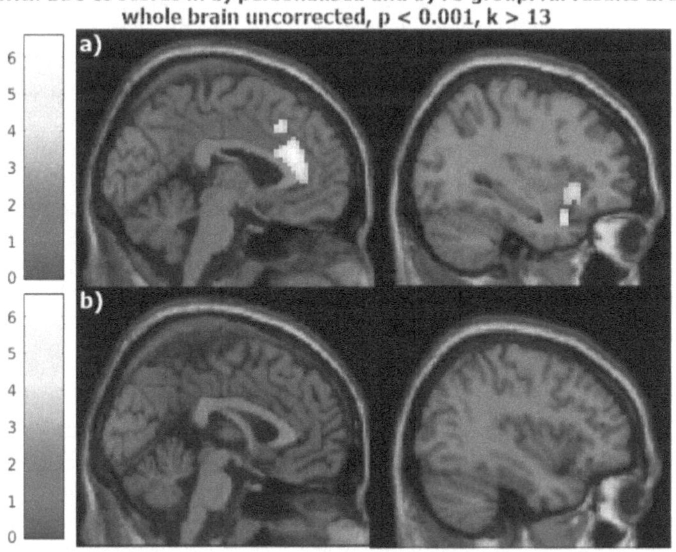

Figure 7: The results of simple linear regression run using SPM12. The personalized iTBS group shows a negative relationship between the FC of DMN with rACC, dACC, and left anterior insula and the BDI-II scores. The F3 iTBS group does not show these trends. Colorbars represents the t-value.

Relationship between DMN and BDI-II scores

 To evaluate the relationship between depression severity and DMN, we ran a linear regression using SPM12. As we are evaluating the DMN, we report results for the BDI-II self-rated scale (p-value < 0.001, k > 13, whole brain uncorrected). In the personalized group we found that the FC of the DMN to rACC, parts of dACC, and left anterior insula (AI) show a negative relationship to the BDI-II scores (Figure 7a). No clusters can be detected for the F3 iTBS group (Figure7b). The FC shows a similar relationship with HAM-D and MADRS scores (data not shown), indicating the robustness of this relationship.

ECG

Personalized iTBS

Comparison of slope across real and sham iTBS condition revealed that personalized real iTBS caused larger deceleration in HR compared to the personalized sham iTBS. This was observed for durations lasting 30s (F = 39.15, p-value = .000**), 45s (F = 48.95, p-value = .000**), and 60s (F = 47.04, p-value = .000**) (Figure 8a). For longer durations (180s, 270s, 360s, and full length) the slope of the RR interval curve did not differ significantly between the real and sham iTBS conditions.

Pursuant to literature, we calculated the RR interval for periods longer than 2 minutes, as a shorter period increases unreliability of RR interval appraisal (51). We therefore compared the mean RR interval for durations only longer than two minutes. We find a trend of higher mean RR interval during personalized real iTBS as compared to sham iTBS for all durations evaluated: 180s (F = 5.099, p-value = 0.024), 270s (F = 5.42, p-value = 0.020), 360s (F = 5.004, p-value = 0.026), full length (F = 5.258, p-value = 0.022) (Figure 8b). These comparisons do not survive multiple comparison correction. The natural logarithm of RMSSD and SDNN also show higher values during real iTBS condition compared to the sham iTBS condition. For these HRV parameters, we also evaluated 60s period as previous works have shown that ECG of 60 seconds duration is enough for reliable and accurate estimation of RMSSD and SDNN parameters (52–54). The ANOVA results comparing the RMSSD and SDNN for real and sham conditions are detailed as follows: RMSSD: 60s (F = 8.320, p-value = 0.004**); 180s (F = 8.098, p-value = 0.005**); 270s (F = 7.892, p-value = 0.005**); 360s (F = 7.308, p-value = 0.007); full length (F = 6.699, p-value = 0.010) (Figure 8c). SDNN: 60s (F = 17.441, p-value = .000**); 180s (F = 10.987, p-value = 0.001**); 270s (F = 8.043, p-value = 0.005**); 360s (F = 7.890, p-value = 0.005**); full length (F = 4.585, p-value = 0.033) (Figure 8d).

F3 iTBS

We conducted ANOVAs comparing the slope, mean RR interval, ln(RMSSD), and ln(SDNN) for real versus sham conditions in the F3 iTBS group. Similar to the personalized iTBS group, the slope of the RR interval curve was significantly higher during the first 30s (F = 12.078, p-value = 0.001**) as well as 45s (F = 9.155, p-value = 0.003**) of real iTBS than the sham iTBS condition. However, unlike the personalized iTBS group, this difference is not significant for the 60s duration (F = 2.466, p-value = 0.117) (Figure 8a). The mean RR interval did not show any significant difference between the real and the sham iTBS conditions in F3 group. The resulting statistics from the ANOVAs are reported as follows: 180s (F = 1.046, p-value = 0.307); 270s (F = 0.984, p-value = 0.322); 360s (F = 0.825, p-value = 0.364); and full-length (F = 1.113, p-value = 0.292) (Figure 8b). The natural logarithm of RMSSD and SDNN also did not show any differences between real and sham F3 iTBS conditions for either of the evaluated

ECG durations [RMSSD: 60s (F = 2.264, p-value = 0.133); 180s (F = 1.737, p-value = 0.188); 270s (F = 2.058, p-value = 0.152); 360s (F = 2.251, p-value = 0.134); full length (F = 2.163, p-value = 0.106) (Figure

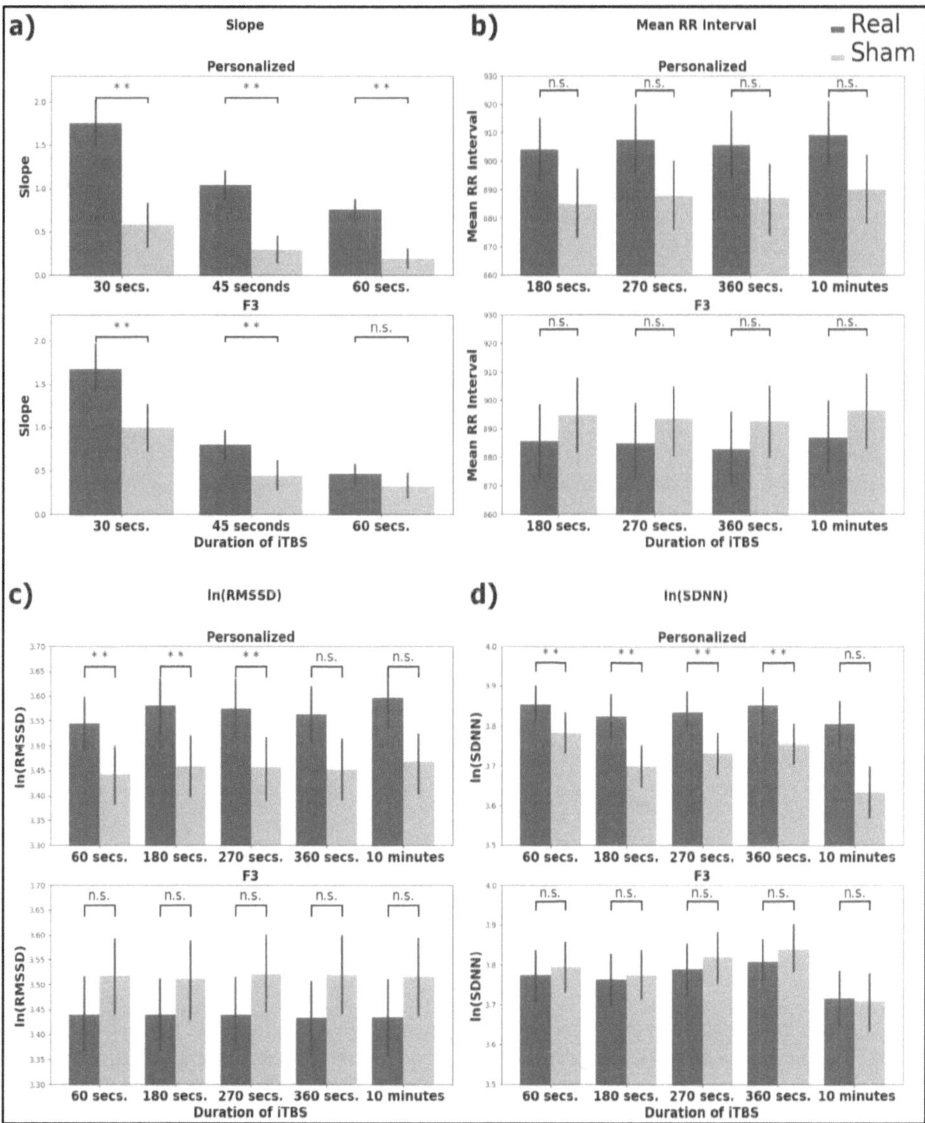

Figure 8: *Changes in HRV in response to personalized (upper panels) and F3 (lower panels) iTBS. a) Slope; b) Mean RR Interval; c) natural logarithm of RMSSD; and d) natural logarithm of the SDNN. [**p-value < 0.05, corrected for multiple comparison]*

8c). SDNN: 60s (F = 0.017, p-value = 0.896); 180s (F = 0.040, p-value = 0.841); 270s (F = 0.206, p-value = 0.650); 360s (F = 0.466, p-value = 0.495); full length (F = 0.547, p-value = 0.460) (Figure 8d)]. Figures 8(a-d) depict the results visually for slope, mean RR interval, natural logarithm of RMSSD and natural logarithm of SDNN, respectively.

Relationship of HRV and sgACC-DMN functional connectivity

As the sgACC showed a decreased FC to the DMN during week 3, we explored the relationships between the FC of sgACC to DMN (quantified by the beta values) and HRV parameters estimated from ECG acquired along with the rsfMRI data. As we had observed differences in the slope of RR interval curve, mean RR intervals, RMSSD, and SDNN, we used these time domain HRV parameters to evaluate their relationship with sgACC-DMN FC. We did not find any relationship between sgACC-DMN FC and mean RR interval, RMSSD or SDNN. However, the slope of the RR interval curve shows a negative relationship with the strength of sgACC-DMN FC. This was specific to the personalized iTBS group only (Figure 9, r = -0.251, p-value = 0.035). This relationship was not observed in the F3 iTBS group (Figure 9, r = -0.066, p-value = 0.592).

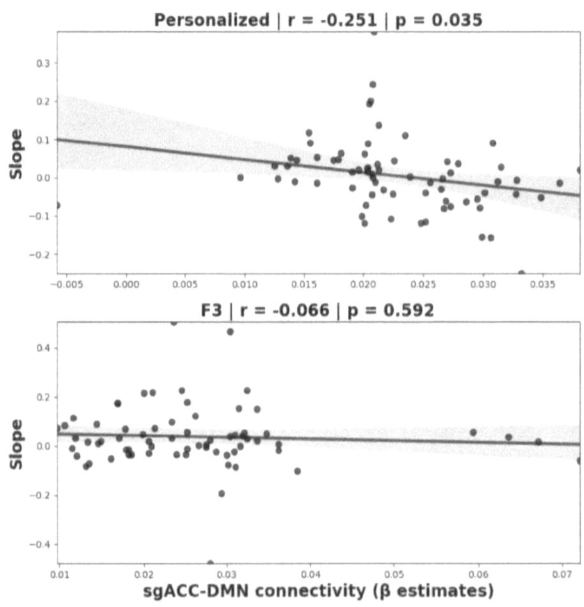

Figure 9: *The scatterplot on the top shows the relationship between the strength of sgACC-DMN FC and its relationship to the slope of the RR interval curve for the entire length of the rsfMRI recording (10 minutes), for the personalized iTBS group. The scatterplots on the bottom is for the F3 iTBS group. A negative relationship is reported only for the case of personalized iTBS group.*

Relationship between HRV and BDI-II scores

A correlation analysis between the relevant time domain HRV parameters (slope, mean RR interval, RMSSD and SDNN) and the BDI-II scores revealed a significant negative relationship for natural logarithm of the RMSSD. This relationship was only observed in the personalized iTBS group (r = -0.379, p-value = 0.001**). The F3 iTBS group did not show any significant relationship between the RMSSD and BDI-II scores (r = 0.146, p-value = 0.231) (Figure 10).

In addition, there was also a similar negative relationship observed between the natural logarithm of SDNN and BDI-II scores. Like RMSSD, this relationship was specifically observed only in the personalized iTBS group (r = -0.393, p-value = 0.000**). The F3 iTBS group did not exhibit this relationship (r = 0.096, p-value = 0.435) (Figure 11).

HA as a predictor of iTBS response

Unlike our results from healthy subjects, we do not see any changes in the FC of the DMN and rACC in patients with depression. We hence evaluated if the changes observed in sgACC-DMN FC correlate to HA scores. We did not observe a relationship between the changes in sgACC-DMN FC and the HA, indicating the functional connectivity changes observed in response to iTBS also did not relate to the HA personality dimension (data not shown). Additionally, we also did not find any evidence for a relationship between the HA scores and therapeutic response resulting from iTBS.

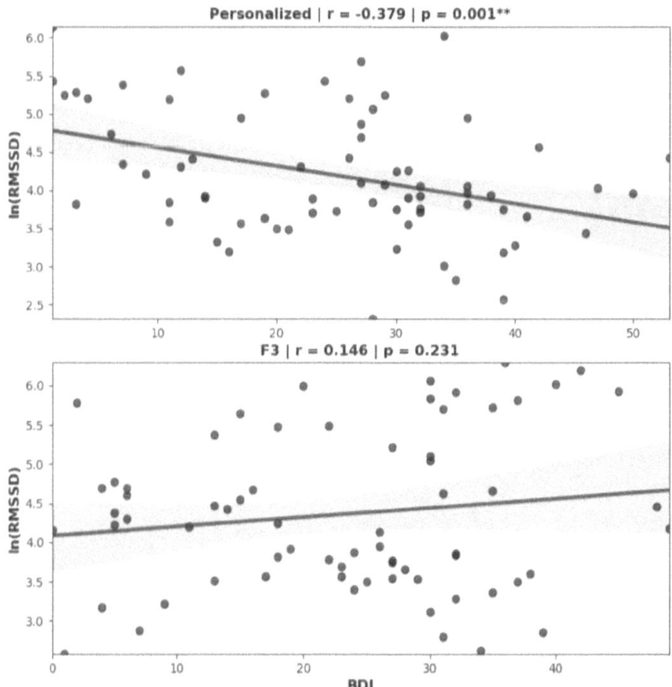

Figure 10: Scatterplots depicting the relationship between the time domain HRV parameter RMSSD and the reported BDI-II scores. A negative relationship is seen for the personalized iTBS groups. However, this is not observed in the F3 iTBS group.

We find that while the patients reported a significantly higher physical sensation from real iTBS, the expected effects on their mental states were the same from both real and sham iTBS. (Supplementary figure 2)

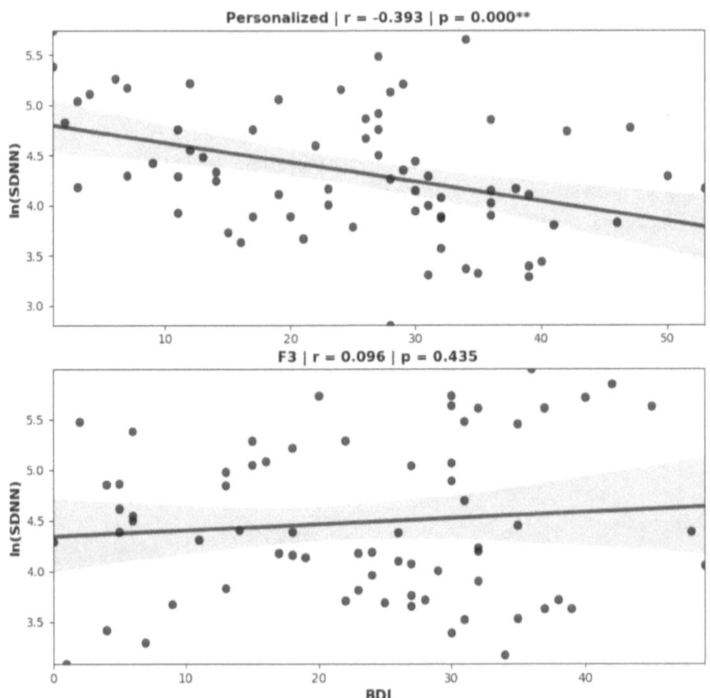

Figure 11: Scatterplots depicting the relationship between the time domain HRV parameter SDNN and the reported BDI-II scores. A negative relationship is seen for the personalized iTBS groups, but not observed in the F3 iTBS group.

Discussion

To the best of our knowledge, we present for the very first time a study comparing real vs. sham personalized and F3 iTBS using multimodal data (rsfMRI, ECG, and behaviour rating scales) in patients with depression. With this novel systems approach, we show that the cluster showing reduced sgACC-DMN connectivity is about four times larger than that resulting from F3 iTBS. We also find significant increase in HRV and a trend-wise decrease in HR in response to real iTBS only for the personalized, but not for F3 iTBS. In light of known relationships between HR and activity in vmPFC (comprising the sgACC) (55,56), we also see a negative relationship between the slope of RR interval time series and sgACC-DMN FC in the personalized iTBS group. In addition, we identified a robust negative correlation between the FC of DMN with the dACC, rACC, and left AI and the BDI-II, HAM-D, MADRS scores exclusively for the personalized iTBS group. Lastly, in support of previous work (30), we observe an inverse relationship between the BDI-II scores and HRV indices. However, we did not find any significant differences in therapeutic response between personalized and F3 iTBS groups in our analysis. This suggests that such intervention driven changes in the FC of the salience network (SN) nodes with the DMN are related to depression severity, which does not necessarily represent the antidepressant response to aiTBS. Although, we identified a trend of larger BDI-II symptom reduction in F3 in comparison to personalized iTBS group, which does not extend to MADRS nor HAMD scales. Nevertheless, clearer effects in the sgACC, extending to HRV, and a relationship between DMN and rating scales are suggestive of more advantageous use of targeted methods such as the personalized approach evaluated here. Previous studies highlighted the importance of the selection of left DLPFC targets based on their FC to sgACC to more effectively mediate antidepressant response (7,8,57). Our preliminary results presented here support their assumption only partially. We hope future analysis of this ongoing prospective study will help to further delineate the effects of personalized and F3 iTBS on treatment response.

Changes in depression rating scales

Here we present results from 38 patients of whom 20 received personalized iTBS and 18 received F3 iTBS. We screened 62 subjects, 37.5% of whom dropped out due to the inability to commit time for the study. Of those who started with the stimulation, 9.1% dropped out from personalized group and 21.74% dropped out from F3 group before finishing the stimulation. The higher dropout rate in F3 group may stem from discomfort associated with the EEG cap used for F3 localisation or lack of treatment response or motivation. Further analysis of distance between optimal location of stimulation and the actual point of stimulation for subjects in F3 group will help clarify if this may explain the higher dropout rate in F3 group. However, in the preliminary analysis both the groups (personalized and F3) showed a significant decrease in depression rating scales at week 6 compared to week 1. Across groups, changes observed in the BDI-II, HAM-D and MADRS rating scales were not significantly different between the two groups, but a trend was seen in the decrease of BDI-II scores, which was higher for the F3 iTBS. This is contrary to our hypothesis that therapeutic response would be higher in the personalized group. But considering that the proportion of subjects who exhibited treatment response (50% or greater reduction in symptom scores based on BDI-II, HAM-D, and MADRS) is comparable and that the larger decrease was seen only in BDI-II scores and not in the others, this finding warrants further scrutiny. One possibility is that this effect is more subjective and therefore arising only when participants rate symptoms themselves. Alternatively, limited sample size can give rise to inconsistent fluctuations in such comparisons due to insufficient power to detect differences across groups.

The therapeutic response is influenced and predicted by many factors related to both the individual's physiology and disease characteristics. One reason for equitable therapeutic response in both groups may be a greater proportion of fast responders in F3 group. Larger studies with neuromodulation or pharmacological treatment have reported presence of clinical response trajectories that broadly classifies patients into groups based on their response to antidepressant treatment. Larsen et al. (58) conducted a trajectory analysis of 1357 patients with depression receiving either escitalopram or placebo over a course of 8 weeks. They reported presence of three classes of responders: fast, slow, and non-responders. Kaster et al. (59) conducted a similar response trajectory analysis on the THREE-D (6) sample for patients receiving either 10 Hz rTMS or iTBS. They reported the presence of four trajectories ranging from non-response to response dependent on baseline depression severity. Given our modest sample sizes, a higher proportion of fast responders in the F3 group could have resulted in a more favourable response seen in this group. While our small sample size prevents any insightful trajectory analysis, future analysis with larger sample should test these assumptions. Another crucial point to note is that Larsen et al. (58) and Kaster et al. (59), report from samples in which antidepressant intervention was applied over a period of 4-8 weeks. While such response trajectory-based subgroups may exist in patients with depression, the use of accelerated protocols might significantly alter these trajectories. Hence direct comparisons must be made cautiously.

In addition to response trajectories, studies have focused on finding predictors of therapeutic response as they might have an inherent clinical value. Knowing which type of patient would present higher chance to respond could reduce the burden of unhelpful treatments, reducing risks and suffering for patients and relatives as well as optimizing team and resources for the clinics. Personality factors have been suggested to predict antidepressant response (22). Baeken et al. (24) delivered 10 Hz rTMS in 36 patients with depression and found that self-directedness (SD; a personality dimension in the TCI) positively predicted the outcome from stimulation. Siddiqi et al. (25) applied 30 sessions of 10 Hz rTMS in 19 patients with depression and reported that persistence (P; another personality dimension in the TCI) positively predicted rTMS treatment outcome. They also found a negative relationship between the SD trait and depression severity. Kopala-Sibley et al. (23) applied 10 Hz rTMS in patients with depression. They evaluated personality using the big five inventory, and measured neuroticism, a trait closely related to HA. A decrease in neuroticism in response to rTMS treatment was reported, but they did not find it predictive of symptom improvement. Based on our work with healthy volunteers (10,11), we had expected that HA personality scores would be predictive of the response to iTBS. However, we do not find evidence in support of our hypothesis. Even though preliminary, if existent, such positive correlation in the DMN decoupling and HA score evidenced in our previous study would have been expected to be seen in this analysis with the given sample size. The literature on predicting rTMS response using personality trait in patients with depression is scarce, but previous works suggests it to be a promising measure for predicting rTMS response (23–25).

Certain classes of medications can influence the antidepressant response of patients receiving rTMS therapy. Hunter et al. (60) reported a positive effect of psychostimulants use and a negative influence of benzodiazepines use on rTMS response. As they argue, the former is likely resulting from influence of psychostimulants on cortical plasticity mediated via their action on the adrenergic pathways (60). The benzodiazepines, on the other hand, influence the GABA levels (60). In patients with depression, magnetic resonance spectroscopy studies have shown lower levels of GABA in the occipital cortex, dorsal prefrontal cortex and ACC (61). Thus, potential upregulation of GABA by rTMS antidepressant treatment is possibly counteracted by benzodiazepines, eventually mitigating the rTMS effects (60). While, it is important to highlight that these relationships are observed in samples receiving non-accelerated rTMS but not iTBS, these studies imply that different classes of pharmacological antidepressants differently influence rTMS outcome. On the other hand, Fitzgerald et al. (62) recently reported that their study sample of patients with depression receiving either accelerated 10 Hz rTMS or iTBS did not show a reduced response associated with benzodiazepine intake. However, they acknowledge the limitations of their results, as they coded the consumption of benzodiazepine merely as a yes/no response. Thus, inaccurate quantification of medication dose could have resulted in lack of effects seen in their sample.

In order to control the influence of medication on iTBS effects, we utilized standardized blood screen panels weekly to measure the medication concentration in the blood stream. This allows more accurate quantification of medication levels that reach the CNS than only registering the dose of intake. In our study, only a proportion of patients (68%, (75% in personalized and 61% in F3), see supplementary table 2 for details on medication load) were under the influence of stable therapeutic ranges of medication, despite no change in type or dose of medication during the course of the study. (see exclusion criteria based on participants' reports). Therefore, in part, therapeutic response across F3 and personalized iTBS groups is likely a result of an interaction between both medication and the aiTBS. Our decision to allow for constant medication use in this trial was based on real world situations, which are more common in the clinical setting. Studies that enrol only medication free participants cannot extrapolate their findings to common patient cohorts, consisting of patients who only partially respond to medication and receive rTMS as an add-on therapy. It is clear that on the one hand, there may be an increase in variability considering such cases; on the other hand, rTMS as add-on treatment to medication may act synergistically as relevant circuitries are already prompted to be modulated for each case, thus increasing specificity. Another possibility is that likely differences in the underlying characteristics of the disease itself can render the symptoms more or less likely to respond to rTMS. Thus, while medication may likely influence symptom improvement in conjunction with effects from iTBS, a diverse disease pathophysiology could also lead patients to respond differently. To accurately delineate effects of medication from other aspects of disease pathophysiology future large studies should continue to incorporate the precise measurements of medication dose and central effects, so that they further unravel the relevant mechanisms of different classes as well as control for potential pharmacological confounders. This will pave way for a better understanding of these subtle, yet complex interaction effects. Delineation of such effects and pathways of interaction will also help improve response to neuromodulation by either discontinuing medication that might mitigate its effects or prescribing medications that might enhance its effects.

Finally, additional factors have been shown to predict rTMS antidepressant outcomes. Both Lisanby et al. (63) and Brakemeier et al. (64) reported that lower levels of medication refractoriness predict better rTMS treatment outcome. Brakemeier et al. (64) also reported that reduction in sleep disturbances are an important predictor of who respond to rTMS. Apart from this, shorter duration of current episode was associated with better rTMS outcome (63), absence of anxiety disorder can positively relate to better rTMS response (63), depression severity may negatively associate with better rTMS outcome (65,66), and also age (67,68) can influence rTMS outcomes. Future work should further explore these factors in detail to account for predictors that truly reveal therapeutic advantage in personalized over F3 iTBS, if any.

Next, by administering PANAS at the beginning and the end of each stimulation day, we also evaluated the changes in PANAS scores to determine if aiTBS can influence short term mood. We report that while the positive affect does not change over the course of the stimulation sessions daily, the negative affect decreases significantly. In patients with depression, one study reported acute changes in mood after rTMS treatment (69). In line with this report, we find a reduction in negative affect in our patient cohort. However, this decrease in negative affect is not limited to real iTBS, and is observable in sham iTBS too, in both the personalized and F3 iTBS groups. We argue that the improvement observed in mood is not a consequence of iTBS but is an effect arising out from the context of receiving treatment and related to the subjects' internal expectations associated with it. In line with this, the subjects reported different scalp sensations corresponding to real and sham iTBS.

However, the expected outcome on their mental states was not different from either real or sham iTBS. Therefore, our results support the idea that differences in the negative PANAS is likely a placebo effect. Also, it supports the notion that the study blinding was effective. Placebo effects are frequently observed in the field of brain stimulation. The elaborate setup employed for both real and sham conditions have the potential to minimize the differences between them (70). In addition, we used an advanced placebo setup that was designed to partially mimic the skin sensation as observed during real stimulation. The significant difference between scalp sensations reported for real and sham iTBS indicates that the placebo setup fails to truly replicate the real iTBS effects. However, the identical procedure of placing the electrodes, placing coils monitored by neuronavigation, and measuring the RMT explain why the subjects may have expected similar outcomes from either week of stimulation (71). Accordingly, we argue that iTBS does not directly influence short term mood. The observed decrease in negative affect from multiple sessions of iTBS in patients with depression is resultant from non-specific effects of iTBS. Thus, the therapeutic benefits of iTBS seem likely to arise from more long-term changes in brain connectivity.

Changes in the DMN and its relationship to depressive symptoms

The DMN plays an important role in the pathophysiology of MDD as past research shows that the DMN is hyperconnected in patients when compared to healthy controls (15,72–74). Particularly, the pgACC, including the sgACC, thalamus, and the precuneus exhibit higher than normal FC to DMN in depression, which except for the thalamus can be normalized after 5 weeks course of 10 Hz rTMS (15). Our results show (figures 5 & 6) that in personalized group, compared to the baseline (R1), the FC of DMN with sgACC, precuneus, and the PCC reduces approximately a week after the end of aiTBS protocol (R3). A similar comparison in the F3 group shows a reduction only in the precuneus region but not the sgACC or the PCC. A comparison of the DMN immediately after the aiTBS (R2) and approximately one week after the end of iTBS (R3) revealed a reduction in FC of DMN to sgACC and precuneus in both groups. However, the sgACC cluster showing a decrease in sgACC-DMN connectivity was four times larger for the personalized group compared to F3 group, indicative of a stronger effect from personalized aiTBS.

The timeline of these effects is revealing of an aiTBS mechanism. We observe a reduced FC of DMN to sgACC and precuneus during R3. The reduction is not an immediate effect because the same regions do not show a reduced FC during the R1-R2 comparison. This implies that the DMN FC of these regions gradually decreases in response to aiTBS. This effect is unlikely from antidepressant medication since the medication dose was stable during the study, as well as at least two weeks prior to start of the study. This ensured that any changes observed in our data (neuroimaging, HRV, or depression severity) were most likely influenced by iTBS. Previously, Baeken et al. (18) used a similar study design to apply aiTBS in patients with depression. They also reported changes in sgACC connectivity developing in time over a period of three weeks (the same time frame in which we observe a decrease in FC). Unlike our results, they did not find reduction in sgACC-DMN FC, but instead found an increase in sgACC-medial orbitofrontal cortex (mOFC), which might stem from a different analysis procedure that did not particularly focus on the DMN. Several other key dissimilarities between the studies could have amounted to the observed differences in the results. Firstly, unlike our study, they did not include a week-long pause between the two stimulation weeks. Secondly, they recruited only medication free patients with treatment resistance depression. Our sample is more naturalistic and representative of depression as a disorder, considering we recruited subjects irrespective of treatment resistance or medication status. Thirdly, they delivered 20 iTBS sessions in four days as opposed to five days in our study. This could also result in differences in how aiTBS modulates the circuits. Lastly, they utilized the anatomical definition of left DLPFC as a target, as against our more functionally relevant rsfMRI based approach. All these disparities could have resulted in the characteristic FC changes observed in each study. Despite the underlying differences in study designs, it is important to note that both results point towards temporally delayed FC changes.

The F3 iTBS group does not show reduced sgACC-DMN FC in R1-R3 comparison even though both groups show a reduced FC of DMN to sgACC and precuneus in the R2-R3 comparison. However, in R2-R3 comparison of the personalized group, the cluster size of sgACC showing a reduced FC to DMN is four times larger than the sgACC cluster reported in F3. Thus, in support of our hypothesis personalized and precise targeting of left DLPFC seemingly results in larger disengagement of sgACC, while inadequate targeting of left DLPFC in F3 group results in poor response of sgACC region. This larger decrease of sgACC-DMN connectivity in R2-R3 comparison, observed for the personalized group, results in a significantly reduced sgACC-DMN connectivity in R1-R3 comparison, but not for the F3 group.

Other regions showing reduced FC to DMN include the PCC and precuneus. Zhang et al. (75) used graph theory to analyze patients with depression and found that left precuneus/PCC shows higher node centrality in depression, indicating a higher connectivity of these regions. Sundermann et al. (76) reported left precuneus having a higher FC to DMN, while the right precuneus displayed a lower FC to DMN. Liston et al. (15) has also reported a higher connectivity between sgACC, and nodes of the DMN, including the precuneus. Thus, the PCC/precuneus region seem also involved within the pathophysiology of depression. Furthermore, the precuneus/PCC region FC is associated with response to antidepressants. For example, Smith et al. (77) and Li et al. (78) showed a decreased precuneus metabolism in response to citalopram and 10 Hz rTMS, respectively. Another study evaluating the influence of 10 Hz rTMS using rsfMRI showed that rTMS reduced the increased FC of bilateral precuneus to the affective network (79). Using iTBS, Persson et al. (80), showed that stimulating at dorsomedial prefrontal cortex (dmPFC) sites resulted in symptomatic alleviation, which correlated with increased FC between the right precuneus and the dmPFC target sites. The precuneus is generally implicated for processing of episodic memory and encoding of self- relevant information (79). As an integral node of the DMN, the increased FC of precuneus to amygdale observed in patients suggest failure to regulate negative rumination on the self (79). Thus, evidence for decreased precuneus-DMN FC after aiTBS may indicate symptomatic benefits through reduced negative perceptions about oneself. Based on our results, we infer that the FC of precuneus to DMN reduces in response to aiTBS but the rsfMRI based targeting seems more effective than F3 at influencing sgACC-DMN FC. We additionally hypothesize that the reduced FC between DMN-precuneus region accompanies an increased FC between the precuneus and the dmPFC, influencing the SN, as reported elsewhere (80). The increase in FC of precuneus and SN could plausibly allow efficient regulation of rumination, as theorized by Persson et al (80). This further hints at mechanisms behind behavioural clinical benefits of aiTBS. However, further evidence of increased FC between precuneus and SN nodes remains to be established through future work.

Several studies in the past have suggested the utility of prospectively selecting left DLPFC targets based on its anticorrelation to sgACC region. These studies retrospectively show that the therapeutic response is proportional to the extent of anticorrelation between the stimulation targets and sgACC

(7,8,57). In support of such findings, our results further build a case for prospectively selecting left DLPFC targets based on their anticorrelation to sgACC for more robust effects in sgACC. Other brain regions have been related to clinical improvement from rTMS. One study of patients with depression receiving 10 Hz rTMS treatment (81) found that rTMS induced decrease in the FC between the left DLPFC stimulation sites and the sgACC was related to reduction in HAM-D scores. Similarly, Salomons et al. (82) also reported association between reduction of these scores and the connectivity of sgACC-DLPFC, sgACC-Insula, sgACC-amygdala, dmPFC-sgACC, dmPFC-thalamus, dmPFC-putamen. In case of iTBS, Persson et al. (80) reported that baseline FC in the dmPFC-precuneus, the sgACC-precuneus and changes in precuneus-dmPFC FC were associated with reduction in MADRS scores. Using aiTBS protocol, Baeken et al. (18) reported a relationship between sgACC-mOFC and hopelessness. However, in our results such relationships between symptom reduction and DMN FC changes were not observed in the sgACC or precuneus. It is important to note that most of analyses investigating biomarkers of clinical improvement after rTMS generally focus on the FC between the stimulation region (dmPFC or DLPFC) and deeper regions like the precuneus, sgACC, thalamus, and putamen, among others. We however analysed the DMN FC, which may explain the lack of such relationships in our analysis. Moreover, when searching for biomarkers of treatment response, the true relationship between a set of sub-symptoms and biomarkers may be masked by inclusion of other non-relevant sub-symptoms (equivalent to addition of noise to true signal). Fried et al. (83) reviewed studies on depression that have analysed sub-symptoms of depression. Their findings suggest that use of sum-total of depressive scores is not conducive of developing insights into the pathophysiology and its interaction with treatment modalities. Notably, there is also considerable heterogeneity in the rTMS protocols used by various groups reporting on relationships between FC and clinical symptoms. Disregarding the variance in rTMS protocols as well as use of partial or summed symptom scores, the lack of relationship between clinical improvement and DMN FC of sgACC or precuneus regions may stem from limited power of our sample. Future work with larger sample size and exploration of symptom subtypes may facilitate fresh insights into the relationship between therapeutic response and DMN FC changes.

We report that DMN FC to the dACC, and the left AI correlates negatively to BDI-II, MADRS, and HAM-D scores only in personalized iTBS. Considering the imprecise nature of F3 iTBS, this may be the reason why these effects are not evidenced in the other group. Despite the limited sample size, the precise delivery of iTBS could increase homogeneity in the effects owing to more targeted engagement of brain regions in the personalized group. Note that these regions are typically considered part of the SN (84–86), which plays an important role in switching between DMN and task positive networks (e.g. CEN). The FC of the DMN with these SN nodes seem to mediate the severity of depressive symptoms. In patients with depression, Manoliu et al. (87) also reported an inverse relationship between the symptom severity and rAI-SN FC. The AI plays an important role in integrating social and affective information (88). Thus, reduced rAI-SN FC results in poor processing of emotional and social cues resulting in failure to relegate between DMN and CEN functioning. The poor switching between DMN and CEN functioning potentially leads to inabilities to divert from negative emotions as observed in ruminative thoughts and reported in (87). The dACC instead plays a more crucial role in conflict resolution and cognitive control (88). A higher dACC-DMN FC may also allow greater regulation of self-directed processes like rumination (80), which may translate into reduced depression severity, as observed in our results. Thus, our results further strengthen the notion that SN nodes play a role in mediating the symptom severity by modulating the communication between other relevant networks such as DMN and CEN.

On a final note, we had hypothesized dACC, rACC, and rAI regions to show a change in FC based on our previous work in healthy samples (11). However, we did not find any statistically significant changes in the FC of the DMN with these regions. Further analysis with larger sample sizes will help clarify if the FC of these regions respond to iTBS, however it must be kept in mind that mechanisms to multiple session iTBS in patients with depression may not follow the same changes as those reported from a single session of iTBS in a healthy cohort. Additionally, we did not find any relationship between the FC of these SN nodes and DMN and that with symptom improvement. This might further imply that these regions may only play a role in mediating symptom severity but not symptom improvement.

Heart rate variability and iTBS in depression – its relationship to rsfMRI and depression severity

Antidepressant treatment with rTMS is associated with improvements in HRV. Udupa et al. (89) applied 15 Hz rTMS in patients with depression over a span of two weeks and found an increase in SDNN and RMSSD in response to rTMS. The increase in SDNN and RMSSD was also greater in the group receiving rTMS compared to the one receiving escitalopram, albeit only the SDNN showed a significantly higher value (89). However, the study did not involve a sham arm. More recently, Iseger et al. (37) showed that patients with depression receiving 3-minute iTBS showed higher HRV during real iTBS as compared to sham iTBS. They also found that within the first 30 seconds of iTBS, the slope of RR interval time series is higher during real stimulation compared to sham. Based on this, we had hypothesized that the HRV parameters would be higher in real iTBS condition compared to sham and even higher for personalized than F3 group. Confirming our hypothesis, in the personalized group we see an increase in HRV and a trend towards reduction of HR after iTBS, the latter not surviving Bonferroni correction. In the personalized group, we find that the slope of the RR interval time series is significantly higher during the first 30 seconds, 45 seconds, and 60 seconds of real iTBS condition compared to sham iTBS. A similar effect is observable in the F3 iTBS group but only during the first 30 seconds and 45 seconds of real iTBS. In line with a previous meta-analysis evaluating the effects of non-invasive brain stimulation on HRV, we report that, in personalized iTBS group, the mean RR interval, the RMSSD, and the SDNN are higher during real iTBS condition, compared to the sham iTBS condition. We observe this effect for multiple time windows of the iTBS stimulation. For the mean RR interval, we used a minimum window of about approximately 3 minutes, as ECG data length shorter than two minutes does not lead to reliable estimate of RR intervals (51). However, for RMSSD and SDNN, ECG duration of 60 seconds suffices for a reliable estimation of these parameters (52–54). The observed HR deceleration within the first 60 seconds indicate that the effects of iTBS are observable immediately from the start of personalized iTBS. While we do see some HR deceleration from F3 iTBS, we do not observe differences in HRV from F3 iTBS. The probable reason for lack of such effects may be that F3 based iTBS does not consider either structural or functional properties of DLPFC region.

Of note, Iseger et al. (90) and Kaur et al. (91) both show changes in HRV parameters in response to F3 rTMS, albeit in healthy subjects. We argue that healthy subjects have a robust ANS, leading to changes in HRV despite more imprecise delivery of stimulation. Contrarily, patients with depression have a lower HRV compared to healthy controls (30), indicative of a poorly functioning ANS. Hence, precise and targeted approaches may be more suitable to elicit HRV changes in response to rTMS, as reported here and in (37). Thus, we argue that we see the modulatory effects on HRV only from personalized iTBS as a result of more targeted iTBS delivered to relevant left DLPFC nodes that were more likely to result in sgACC effects across the frontal vagal pathway (27,91). This is further supported by our imaging results, which show that the personalized approach results in more robust reduction compared to F3, in the sgACC-DMN FC during R2-R3 comparison.

Considering personalized iTBS affects sgACC-DMN more robustly, it is important to note that sgACC is also involved in the autonomic aspects of emotional expressions (28) as well as cardiac control. One study recently showed a covariance between the cardiac-cycle duration (RR interval) and the neuronal firing rate within the mid- to anterior-cingulate cortex (55). Although, their results are limited to only 11% of cells recorded from the ACC regions, it nevertheless highlights a relationship between activity in the ACC and RR interval. Another study using magnetoencephalography (MEG) recordings in healthy volunteers, elegantly depicted a covariance between the heartbeat-evoked response in vmPFC and thoughts related to oneself (56). The heart-evoked responses were defined as the MEG signal recorded between two heart beats. Their work depicted a direct link between the idea of selfhood, interoceptive thoughts and the activity within the DMN (56). In line with the known role of sgACC in cardiac functioning, we report a negative relationship between the slope of the RR interval time series and the sgACC-DMN FC. This implies that the lower the sgACC-DMN FC, the higher the tendency for the HR to change (positive slope depicts HR decrease while a negative slope indicates HR increase). Recent work with marmosets has shown that over-activity within the sgACC is causally linked to reduced parasympathetic activity (increased heart rate) (92). We find that a general trend of positive slope (decreasing heart rate) is associated with lower sgACC-DMN FC, while more negative slope (increasing heart rate) is associated with stronger sgACC-DMN FC. Our results provide additional evidence for a relationship between autonomic functioning and the sgACC-DMN FC and reinforce the dual purpose of DMN as a network for self-referential processing (93) and for autonomic regulation (28,94).

The proposed relationship of the FC of DMN with depression severity as well as autonomic functioning indicates a possible relationship between autonomic functioning and depression severity. A previous meta-analysis reported a negative relationship between depression severity and HRV parameters (30). Agelink et al. (95,96) also reported a negative relationship between the depression severity and the HRV indices including the RMSSD. Yeh et al. (97) studying the effects of two different pharmacological antidepressants in patients with depression also reported a negative relationship between the depression severity and multiple indices of HRV. Upholding the previous findings, we report that the natural logarithm of the RMSSD and the SDNN show a negative correlation to the BDI-II scores only in the personalized iTBS group. This indicates with more severe depression the reported HRV is lower as compared to less severe depression. While rating scale scores are negatively correlated to HRV, they also correlated negatively to the FC of DMN with dACC, rACC and the left AI. Previous work shows that the FC of dACC with the thalamus, cingulate cortex and the amygdale , as well as the FC of amygdale and anterior insula show a positive relationship with HRV indices (98). In addition, Oppenheimer et al. also reported significant effects from left insula stimulation on mean RR intervals (99). Considering the suggested role of the SN nodes in autonomic regulation, our evidence also implies a possible relationship between HRV and the FC of SN nodes and DMN.

Limitations

The foremost limitation of the study remains the sample size arising from evaluation of our sample differently from a cross-over design. A cross-over design allows detecting treatment effects even with smaller sample sizes, with subjects acting as suitable controls for themselves. However, the effects from real iTBS manifested over the pause week after the end of stimulation. The strong carryover effects from first week of stimulation make correction for sham effects using the second week of stimulation inaccurate. This prompted us using a different analysis design, resulting in reduction in sample size. While each group had approximately 20 subjects, the sample size receiving either real or sham iTBS is only n ~ 10 in this preliminary analysis. Thus, the results must be interpreted with caution. Small sample sizes are liable to considerable influences from underlying variability of subjects. Large sample sizes allow averaging out such nuances thus allowing better delineation of signal from the noise. While underlying variability could have influenced the results, especially so for F3 iTBS group, it can be said that most of the reported results will likely uphold with larger sample size. In addition, lack of differences in symptomatic improvement between personalized and F3 group could also mean that the impact of more precise FC changes may not be captured by the sum-total from the symptom scales of BDI-II, HAM-D, or MADRS. Future analysis should explore specific changes in symptom clusters to further delineate the behavioural effects of personalized versus F3 iTBS.

Conclusions

Our study shows that compared to F3 iTBS, personalized iTBS is more robust in reducing the sgACC-DMN FC as well as increasing HRV and reducing HR. We further report a negative relationship between depression severity and HRV indices. In addition, severity of depression correlated negatively with the FC of dACC, rACC, and the left AI, indicating a role of the SN nodes in mediating perception of externally- and self-reported symptoms. In the F3 group, the lack of a relationship between the symptom scores and the DMN FC as well as between BDI-II scores and the HRV parameters, and the absence of HRV modulatory effects is noteworthy. We argue that this is in fact indicative of better targeting methods employed in personalized iTBS group, culminating in more pronounced and homogenous effects across the neuroimaging and ECG data. Although these results further strengthen the case for adapting personalized techniques in clinical practice, an evident difference between treatment responses across groups has not yet been observed.

References

1. Lefaucheur J-P, Aleman A, Baeken C, Benninger DH, Brunelin J, Di Lazzaro V, et al. Evidence-based guidelines on the therapeutic use of repetitive transcranial magnetic stimulation (rTMS): An update (2014–2018). Clin Neurophysiol. 2020 Feb;131(2):474–528.

2. Bakker N, Shahab S, Giacobbe P, Blumberger DM, Daskalakis ZJ, Kennedy SH, et al. rTMS of the Dorsomedial Prefrontal Cortex for Major Depression: Safety, Tolerability, Effectiveness, and Outcome Predictors for 10 Hz Versus Intermittent Theta-burst Stimulation. Brain Stimul. 2015 Mar 3;8(2):208–

3.   Duprat R, Desmyter S, Rudi DR, van Heeringen K, Van den Abbeele D, Tandt H, et al. Accelerated intermittent theta burst stimulation treatment in medication-resistant major depression: A fast road to remission? J Affect Disord. 2016 Aug;200:6–14.

4.   Bulteau S, Sébille V, Fayet G, Thomas-Ollivier V, Deschamps T, Bonnin-Rivalland A, et al. Efficacy of intermittent Theta Burst Stimulation (iTBS) and 10-Hz high-frequency repetitive transcranial magnetic stimulation (rTMS) in treatment-resistant unipolar depression: study protocol for a randomised controlled trial. Trials. 2017 Dec 13;18(1):17.

5.   Li C-T, Chen M-H, Juan C-H, Huang H-H, Chen L-F, Hsieh J-C, et al. Efficacy of prefrontal theta-burst stimulation in refractory depression: a randomized sham-controlled study. Brain. 2014 Jul;137(7):2088–98.

6.   Blumberger DM, Vila-Rodriguez F, Thorpe KE, Feffer K, Noda Y, Giacobbe P, et al. Effectiveness of theta burst versus high-frequency repetitive transcranial magnetic stimulation in patients with depression (THREE-D): a randomised non-inferiority trial. Lancet. 2018 Apr 28;391(10131):1683–92.

7.   Cash RFH, Zalesky A, Thomson RH, Tian Y, Cocchi L, Fitzgerald PB. Subgenual Functional Connectivity Predicts Antidepressant Treatment Response to Transcranial Magnetic Stimulation: Independent Validation and Evaluation of Personalization. Biol Psychiatry. 2019;2018–20.

8.   Fox MD, Liu H, Pascual-Leone A. Identification of reproducible individualized targets for treatment of depression with TMS based on intrinsic connectivity. Neuroimage. 2013;66:151–60.

9.   Weigand A, Horn A, Caballero R, Cooke D, Stern AP, Taylor SF, et al. Prospective Validation That Subgenual Connectivity Predicts Antidepressant Efficacy of Transcranial Magnetic Stimulation Sites. Biol Psychiatry. 2018 Jul 1;84(1):28–37.

10.   Singh A, Erwin-Grabner T, Sutcliffe G, Antal A, Paulus W, Goya-Maldonado R. Personalized repetitive transcranial magnetic stimulation temporarily alters default mode network in healthy subjects. Sci Rep. 2019 Dec 4;9(1):5631.

11.   Singh A, Erwin-Grabner T, Sutcliffe G, Paulus W, Dechent P, Antal A, et al. Default mode network alterations after intermittent theta burst stimulation in healthy subjects. Transl Psychiatry. 2020;10(1):75.

12.   Cole EJ, Stimpson KH, Bentzley BS, Gulser M, Cherian K, Tischler C, et al. Stanford Accelerated Intelligent Neuromodulation Therapy for Treatment-Resistant Depression. Am J Psychiatry. 2020 Apr 7;12(2):1–11.

13.   Siddiqi SH, Trapp NT, Hacker CD, Laumann TO, Kandala S, Hong X, et al. Repetitive Transcranial Magnetic Stimulation with Resting-State Network Targeting for Treatment-Resistant Depression in Traumatic Brain Injury: A Randomized, Controlled, Double-Blinded Pilot Study. J Neurotrauma. 2019 Apr 15;36(8):1361–74.

14.   Trapp NT, Bruss J, King Johnson M, Uitermarkt BD, Garrett L, Heinzerling A, et al. Reliability of targeting methods in TMS for depression: Beam F3 vs. 5.5 cm. Brain Stimul. 2020;13(3):578–81.

15.   Liston C, Chen AC, Zebley BD, Drysdale AT, Gordon R, Leuchter B, et al. Default mode network mechanisms of transcranial magnetic stimulation in depression. Biol Psychiatry. 2014 Oct 1;76(7):517–26.

16.   Taylor SF, Ho SS, Abagis T, Angstadt M, Maixner DF, Welsh RC, et al. Changes in brain connectivity during a sham-controlled, transcranial magnetic stimulation trial for depression. J Affect Disord. 2018;232:143–51.

17.   Mayberg HS. Targeted electrode-based modulation of neural circuits for depression. J Clin Invest. 2009 Apr 1;119(4):717–25.

18.   Baeken C, Duprat R, Wu G-R, De Raedt R, van Heeringen K. Subgenual Anterior Cingulate–Medial Orbitofrontal Functional Connectivity in Medication-Resistant Major Depression: A Neurobiological Marker for Accelerated Intermittent Theta Burst Stimulation Treatment? Biol Psychiatry Cogn Neurosci Neuroimaging. 2017 Oct 1;2(7):556–65.

19.   Duprat R, Wu G-R, De Raedt R, Baeken C. Accelerated iTBS treatment in depressed patients differentially modulates reward system activity based on anhedonia. World J Biol Psychiatry. 2018 Oct 3;19(7):497–508.

20.   Krepel N, Rush AJ, Iseger TA, Sack AT, Arns M. Can psychological features predict antidepressant response to rTMS? A Discovery–Replication approach. Psychol Med. 2019 Jan 24;1–9.

21.   Balestri M, Porcelli S, Souery D, Kasper S, Dikeos D, Ferentinos P, et al. Temperament and character influence on depression treatment outcome. J Affect Disord. 2019 Jun;252:464–74.

22.   Kampman O, Poutanen O, Illi A, Setälä-Soikkeli E, Viikki M, Nuolivirta T, et al. Temperament profiles, major depression, and response to treatment with SSRIs in psychiatric outpatients. Eur Psychiatry. 2012 May 1;27(4):245–9.

23.   Kopala-Sibley DC, Chartier GB, Bhanot S, Cole J, Chan PY, Berlim MT, et al. Personality Trait Predictive Utility and Stability in Transcranial Magnetic Stimulation (rTMS) for Major Depression: Dissociation of Neuroticism and Self-Criticism. Can J Psychiatry. 2020 Apr 1;65(4):264–72.

24.   Baeken C, Desmyter S, Duprat R, De Raedt R, Van denabbeele D, Tandt H, et al. Self-directedness: An indicator for clinical response to the HF-rTMS treatment in refractory melancholic depression. Psychiatry Res. 2014 Dec 15;220(1–2):269–74.

25.   Siddiqi SH, Chockalingam R, Cloninger CR, Lenze EJ, Cristancho P. Use of the temperament and character inventory to predict response to repetitive transcranial magnetic stimulation for major depression. J Psychiatr Pract. 2016;22(3):193–202.

26.   Jain FA, Cook IA, Leuchter AF, Hunter AM, Davydov DM, Ottaviani C, et al. Heart rate variability and treatment outcome in major depression: A pilot study. Int J Psychophysiol. 2014;93(2):204–10.

27.   Iseger TA, van Bueren NER, Kenemans JL, Gevirtz R, Arns M. A frontal-vagal network theory for Major Depressive Disorder: Implications for optimizing neuromodulation techniques. Brain Stimul. 2020 Jan 1;13(1):1–9.

28.   Thayer JF, Åhs F, Fredrikson M, Sollers JJ, Wager TD. A meta-analysis of heart rate variability and neuroimaging studies: Implications for heart rate variability as a marker of stress and health. Neurosci Biobehav Rev. 2012;36(2):747–56.

29.   Benarroch EE. Central Autonomic Control. Prim Auton Nerv Syst. 2012;9–12.

30.   Kemp AH, Quintana DS, Gray MA, Felmingham KL, Brown K, Gatt JM. Impact of Depression and Antidepressant Treatment on Heart Rate Variability: A Review and Meta-Analysis. Biol Psychiatry. 2010;67(11):1067–74.

31.   Wager TD, Davidson ML, Hughes BL, Lindquist MA, Ochsner KN. Prefrontal-Subcortical Pathways Mediating Successful Emotion Regulation. Neuron. 2008;59(6):1037–50.

32.   Wager TD, Waugh CE, Lindquist M, Noll DC, Fredrickson BL, Taylor SF. Brain mediators of cardiovascular responses to social threat. Part I: Reciprocal dorsal and ventral sub-regions of the medial prefrontal cortex and heart-rate reactivity. Neuroimage. 2009;47(3):821–35.

33.   Wager TD, van Ast VA, Hughes BL, Davidson ML, Lindquist MA, Ochsner KN. Brain mediators of cardiovascular responses to social threat, Part II: Prefrontal-subcortical pathways and relationship with anxiety. Neuroimage. 2009;47(3):836–51.

34.   Ehrenthal JC, Herrmann-Lingen C, Fey M, Schauenburg H. Altered cardiovascular adaptability in

depressed patients without heart disease. World J Biol Psychiatry. 2010 Jan 11;11(3):586–93.

35. Udupa K, Sathyaprabha TN, Thirthalli J, Kishore KR, Lavekar GS, Raju TR, et al. Alteration of cardiac autonomic functions in patients with major depression: A study using heart rate variability measures. J Affect Disord. 2007 Jun 1;100(1–3):137–41.

36. Makovac E, Thayer JF, Ottaviani C. A meta-analysis of non-invasive brain stimulation and autonomic functioning: Implications for brain-heart pathways to cardiovascular disease. Neurosci Biobehav Rev. 2017 Mar 1;74:330–41.

37. Iseger TA, Arns M, Downar J, Blumberger DM, Daskalakis ZJ, Vila-Rodriguez F. Cardiovascular differences between sham and active iTBS related to treatment response in MDD. Brain Stimul. 2020;13(1):167–74.

38. Rossi S, Hallett M, Rossini PM, Pascual-Leone A, Avanzini G, Bestmann S, et al. Safety, ethical considerations, and application guidelines for the use of transcranial magnetic stimulation in clinical practice and research. Clin Neurophysiol. 2009;120(12):2008–39.

39. Oldfield RC. The assessment and analysis of handedness: The Edinburgh inventory. Neuropsychologia. 1971 Mar 1;9(1):97–113.

40. Lehrl S, Triebig G, Fischer B. Multiple choice vocabulary test MWT as a valid and short test to estimate premorbid intelligence. Acta Neurol Scand. 1995 May;91(5):335–45.

41. Wittorf A, Wiedemann G, Klingberg S. Mehrfachwahl-Wortschatz-Intelligenztest MWT bei Schizophrenie: Valides Maß der prämorbiden Intelligenz? Psychiatr Prax. 2013 Oct 2;41(02):95–100.

42. Schulze L, Feffer K, Lozano C, Giacobbe P, Daskalakis ZJ, Blumberger DM, et al. Number of pulses or number of sessions? An open-label study of trajectories of improvement for once-vs. twice-daily dorsomedial prefrontal rTMS in major depression. Brain Stimul. 2017;11:327–36.

43. Moeller S, Yacoub E, Olman CA, Auerbach E, Strupp J, Harel N, et al. Multiband multislice GE-EPI at 7 tesla, with 16-fold acceleration using partial parallel imaging with application to high spatial and temporal whole-brain fMRI. Magn Reson Med. 2010 May 1;63(5):1144–53.

44. Setsompop K, Gagoski BA, Polimeni JR, Witzel T, Wedeen VJ, Wald LL. Blipped-controlled aliasing in parallel imaging for simultaneous multislice echo planar imaging with reduced g-factor penalty. Magn Reson Med. 2012 May 1;67(5):1210–24.

45. Jenkinson M, Beckmann CF, Behrens TEJ, Woolrich MW, Smith SM. FSL. Neuroimage. 2012 Aug 15;62(2):782–90.

46. Virtanen P, Gommers R, Oliphant TE, Haberland M, Reddy T, Cournapeau D, et al. SciPy 1.0: fundamental algorithms for scientific computing in Python. Nat Methods. 2020;17(3):261–72.

47. Lipponen JA, Tarvainen MP. A robust algorithm for heart rate variability time series artefact correction using novel beat classification. J Med Eng Technol. 2019;43(3):173–81.

48. Thayer JF, Lane RD. The role of vagal function in the risk for cardiovascular disease and mortality. Biol Psychol. 2007 Feb;74(2):224–42.

49. Schneider M, Schwerdtfeger A. Autonomic dysfunction in posttraumatic stress disorder indexed by heart rate variability: a meta-analysis. Psychol Med. 2020;1–12.

50. Tik M, Hoffmann A, Sladky R, Tomova L, Hummer A, Navarro de Lara L, et al. Towards understanding rTMS mechanism of action: Stimulation of the DLPFC causes network-specific increase in functional connectivity. Neuroimage. 2017 Nov;162:289–96.

51. Shaffer F, Ginsberg JP. An Overview of Heart Rate Variability Metrics and Norms. Front Public Heal. 2017 Sep 28;5:258.

52.   Chen YS, Clemente FM, Bezerra P, Lu YX. Ultra-short-term and short-term heart rate variability recording during training camps and an international tournament in U-20 national futsal players. Int J Environ Res Public Health. 2020;17(3):1–12.

53.   Salahuddin L, Cho J, Jeong MG, Kim D. Ultra short term analysis of heart rate variability for monitoring mental stress in mobile settings. Annu Int Conf IEEE Eng Med Biol - Proc. 2007;4656–9.

54.   McNames J, Aboy M. Reliability and accuracy of heart rate variability metrics versus ECG segment duration. Med Biol Eng Comput. 2006;44(9):747–56.

55.   Kim K, Ladenbauer J, Babo-Rebelo M, Buot A, Lehongre K, Adam C, et al. Resting-state neural firing rate is linked to cardiac-cycle duration in the human cingulate and parahippocampal cortices. J Neurosci. 2019;39(19):3676–86.

56.   Babo-Rebelo M, Richter CG, Tallon-Baudry C. Neural responses to heartbeats in the default network encode the self in spontaneous thoughts. J Neurosci. 2016;36(30):7829–40.

57.   Fox MD, Buckner RL, White MP, Greicius MD, Pascual-Leone A. Efficacy of transcranial magnetic stimulation targets for depression is related to intrinsic functional connectivity with the subgenual cingulate. Biol Psychiatry. 2012;72(7):595–603.

58.   Larsen KG, Kennedy SH, Reines EH, Thase ME. Patient Response Trajectories in Major Depressive Disorder. Psychopharmacol Bull. 2020 Sep 14;50(4):8–28.

59.   Kaster TS, Downar J, Vila-Rodriguez F, Thorpe KE, Feffer K, Noda Y, et al. Trajectories of response to dorsolateral prefrontal rTMS in major depression: A THREE-D study. Am J Psychiatry. 2019 May 15;176(5):367–75.

60.   Hunter AM, Minzenberg MJ, Cook IA, Krantz DE, Levitt JG, Rotstein NM, et al. Concomitant medication use and clinical outcome of repetitive Transcranial Magnetic Stimulation (rTMS) treatment of Major Depressive Disorder. Brain Behav. 2019 Apr 2;e01275.

61.   Pehrson AL, Sanchez C. Altered γ-aminobutyric acid neurotransmission in major depressive disorder: A critical review of the supporting evidence and the influence of serotonergic antidepressants. Drug Des Devel Ther. 2015;9:603–24.

62.   Fitzgerald PB, Daskalakis ZJ, Hoy KE. Benzodiazepine use and response to repetitive transcranial magnetic stimulation in Major Depressive Disorder. Brain Stimul. 2020;13(3):694–5.

63.   Lisanby SH, Husain MM, Rosenquist PB, Maixner D, Gutierrez R, Krystal A, et al. Daily left prefrontal repetitive transcranial magnetic stimulation in the acute treatment of major depression: Clinical predictors of outcome in a multisite, randomized controlled clinical trial. Neuropsychopharmacology. 2009;34(2):522–34.

64.   Brakemeier EL, Luborzewski A, Danker-Hopfe H, Kathmann N, Bajbouj M. Positive predictors for antidepressive response to prefrontal repetitive transcranial magnetic stimulation (rTMS). J Psychiatr Res. 2007;41(5):395–403.

65.   Kilts CD, Wade AG, Andersen HF, Schlaepfer TE. Baseline severity of depression predicts antidepressant drug response relative to escitalopram. Expert Opin Pharmacother. 2009 Apr 24;10(6):927–36.

66.   Grammer GG, Kuhle AR, Clark CS, Dretsch MN, Williams KA, Cole JT. Severity of Depression Predicts Remission Rates Using Transcranial Magnetic Stimulation. Front Psychiatry. 2015 Sep 1;6:1–5.

67.   Trevizol AP, Downar J, Vila-Rodriguez F, Thorpe KE, Daskalakis ZJ, Blumberger DM. Predictors of remission after repetitive transcranial magnetic stimulation for the treatment of major depressive disorder: An analysis from the randomised non-inferiority THREE-D trial. EClinicalMedicine. 2020 May;22:100349.

68.   Fregni F, Marcolin MA, Myczkowski M, Amiaz R, Hasey G, Rumi DO, et al. Predictors of antidepressant

response in clinical trials of transcranial magnetic stimulation. Int J Neuropsychopharmacol. 2006;9(6):641–54.

69.   Szuba MP, O'Reardon JP, Rai AS, Snyder-Kastenberg J, Amsterdam JD, Gettes DR, et al. Acute mood and thyroid stimulating hormone effects of transcranial magnetic stimulation in major depression. Biol Psychiatry. 2001;50(1):22–7.

70.   Kaptchuk TJ, Goldman P, Stone DA, Stason WB. Do medical devices have enhanced placebo effects? J Clin Epidemiol. 2000;53(8):786–92.

71.   Burke MJ, Kaptchuk TJ, Pascual-Leone A. Challenges of differential placebo effects in contemporary medicine: The example of brain stimulation. Ann Neurol. 2019 Jan 6;85(1):12–20.

72.   Sheline YI, Barch DM, Price JL, Rundle MM, Vaishnavi SN, Snyder AZ, et al. The default mode network and self-referential processes in depression. Proc Natl Acad Sci. 2009 Feb 10;106(6):1942–7.

73.   Greicius MD, Flores BH, Menon V, Glover GH, Solvason HB, Kenna H, et al. Resting-State Functional Connectivity in Major Depression: Abnormally Increased Contributions from Subgenual Cingulate Cortex and Thalamus. Biol Psychiatry. 2007;62(5):429–37.

74.   Cooney RE, Joormann J, Eugène F, Dennis EL, Gotlib IH. Neural correlates of rumination in depression. Cogn Affect Behav Neurosci. 2010 Dec;10(4):470–8.

75.   Zhang J, Wang J, Wu Q, Kuang W, Huang X, He Y, et al. Disrupted Brain Connectivity Networks in Drug-Naive, First-Episode Major Depressive Disorder. Biol Psychiatry. 2011;70(4):334–42.

76.   Sundermann B, Olde Lütke Beverborg M, Pfleiderer B. Toward literature-based feature selection for diagnostic classification: a meta-analysis of resting-state fMRI in depression. Front Hum Neurosci. 2014 Sep 10;8(September):1–12.

77.   Smith GS, Kramer E, Hermann C, Ma Y, Dhawan V, Chaly T, et al. Serotonin Modulation of Cerebral Glucose Metabolism in Depressed Older Adults. Biol Psychiatry. 2009;66(3):259–66.

78.   Li C-T, Chen L-F, Tu P-C, Wang S-J, Chen M-H, Su T-P, et al. Impaired Prefronto-Thalamic Functional Connectivity as a Key Feature of Treatment-Resistant Depression: A Combined MEG, PET and rTMS Study. Paul F, editor. PLoS One. 2013 Aug 2;8(8):e70089.

79.   Chen F, Gu C, Zhai N, Duan H, Zhai A, Zhang X. Repetitive Transcranial Magnetic Stimulation Improves Amygdale Functional Connectivity in Major Depressive Disorder. Front Psychiatry. 2020 Jul 31;11:1.

80.   Persson J, Struckmann W, Gingnell M, Fällmar D, Bodén R. Intermittent theta burst stimulation over the dorsomedial prefrontal cortex modulates resting-state connectivity in depressive patients: A sham-controlled study. Behav Brain Res. 2020;394(July):112834.

81.   Hadas I, Sun Y, Lioumis P, Zomorrodi R, Jones B, Voineskos D, et al. Association of Repetitive Transcranial Magnetic Stimulation Treatment With Subgenual Cingulate Hyperactivity in Patients With Major Depressive Disorder: A Secondary Analysis of a Randomized Clinical Trial. JAMA Netw open. 2019 Jun 5;2(6):e195578.

82.   Salomons T V, Dunlop K, Kennedy SH, Flint A, Geraci J, Giacobbe P, et al. Resting-State Cortico-Thalamic-Striatal Connectivity Predicts Response to Dorsomedial Prefrontal rTMS in Major Depressive Disorder. Neuropsychopharmacology. 2014 Jan 13;39(2):488–98.

83.   Fried EI, Nesse RM. Depression sum-scores don't add up: Why analyzing specific depression symptoms is essential. BMC Med. 2015;13(1):1–11.

84.   Menon V, Uddin LQ. Saliency, switching, attention and control: a network model of insula function. Brain Struct Funct. 2010 Jun 29;214(5–6):655–67.

85.   Markett S, Weber B, Voigt G, Montag C, Felten A, Elger C, et al. Intrinsic connectivity networks and

personality: The temperament dimension harm avoidance moderates functional connectivity in the resting brain. Neuroscience. 2013;240:98–105.

86. Bush G, Luu P, Posner MI. Cognitive and emotional influences in anterior cingulate cortex. Trends Cogn Sci. 2000 Jun 1;4(6):215–22.

87. Manoliu A, Meng C, Brandl F, Doll A, Tahmasian M, Scherr M, et al. Insular dysfunction within the salience network is associated with severity of symptoms and aberrant inter-network connectivity in major depressive disorder. Front Hum Neurosci. 2014;7(January):1–17.

88. Menon V. Large-scale brain networks and psychopathology: a unifying triple network model. Trends Cogn Sci. 2011 Oct 1;15(10):483–506.

89. Udupa K, Sathyaprabha TN, Thirthalli J, Kishore KR, Raju TR, Gangadhar BN. Modulation of cardiac autonomic functions in patients with major depression treated with repetitive transcranial magnetic stimulation. J Affect Disord. 2007 Dec;104(1–3):231–6.

90. Iseger TA, Padberg F, Kenemans JL, Gevirtz R, Arns M. Neuro-Cardiac-Guided TMS (NCG-TMS): Probing DLPFC-sgACC-vagus nerve connectivity using heart rate – First results. Brain Stimul. 2017 Sep;10(5):1006–8.

91. Kaur M, Michael JA, Hoy KE, Fitzgibbon BM, Ross MS, Iseger TA, et al. Investigating high- and low-frequency neuro-cardiac-guided TMS for probing the frontal vagal pathway. Brain Stimul. 2020 May;13(3):931–8.

92. Alexander L, Wood CM, Gaskin PLR, Sawiak SJ, Fryer TD, Hong YT, et al. Over-activation of primate subgenual cingulate cortex enhances the cardiovascular, behavioral and neural responses to threat. Nat Commun. 2020 Dec 26;11(1):5386.

93. Raichle ME. The Brain's Default Mode Network. Annu Rev Neurosci. 2015 Jul 8;38(1):433–47.

94. Beissner F, Meissner K, Bär KJ, Napadow V. The autonomic brain: An activation likelihood estimation meta-analysis for central processing of autonomic function. J Neurosci. 2013;33(25):10503–11.

95. Agelink MW, Boz C, Ullrich H, Andrich J. Relationship between major depression and heart rate variability. Clinical consequences and implications for antidepressive treatment. Psychiatry Res. 2002;113(1–2):139–49.

96. Agelink MW, Majewski T, Wurthmann C, Postert T, Linka T, Rotterdam S, et al. Autonomic neurocardiac function in patients with major depression and effects of antidepressive treatment with nefazodone. J Affect Disord. 2001;62(3):187–98.

97. Yeh TC, Kao LC, Tzeng NS, Kuo TBJ, Huang SY, Chang CC, et al. Heart rate variability in major depressive disorder and after antidepressant treatment with agomelatine and paroxetine: Findings from the Taiwan Study of Depression and Anxiety (TAISDA). Prog Neuro-Psychopharmacology Biol Psychiatry. 2016;64:60–7.

98. Chang C, Metzger CD, Glover GH, Duyn JH, Heinze HJ, Walter M. Association between heart rate variability and fluctuations in resting-state functional connectivity. Neuroimage. 2013;68:93–104.

99. Oppenheimer SM, Gelb A, Girvin JP, Hachinski VC. Cardiovascular effects of human insular cortex stimulation. Neurology. 1992;42(9):1727–32.

7 Summary and future perspectives

A personalized approach for rTMS target selection based on rsfMRI data is suggested as a viable way to overcome variability in treatment response. In chapter 3 the current status of rTMS use in the treatment of major depressive disorder is reviewed. Chapter 4 and 5 present a proof-of-concept demonstration of a protocol developed by our research group, initially applied in healthy individuals to locate the personalized left DLPFC targets for stimulation. Considering the success of this method to yield targets and to robustly modulate the DMN in pgACC regions, we theorize that this method would have great potential for clinical application. In Chapter 6, we then prospectively tested the utility of this personalized approach in a clinical cohort receiving 20 aiTBS sessions and qualitatively compared it against the standard F3 (non-personalized) approach. In a preliminary analysis, we report robust network effects that are aligned with cardiac changes favouring the personalized approach.

The sgACC is an important DMN node implicated in emotion regulation [268]. Disruptions in sgACC functional connectivity underlay depression pathology. The antidepressant response of rTMS also influences the aberrant connectivity of sgACC with other brain regions [146, 160, 161]. However, a proportion of patients still fail to respond to rTMS antidepressant treatment. This has led to a growing need to understand and address the sources of inter-individual variability in rTMS response. Based on previous works, rTMS at left DLPFC targets with stronger anti-correlation to sgACC, result in better rTMS antidepressant response. This suggests certain nodes within the left DLPFC allow more efficient trans-synaptic propagation of rTMS effects to deeper brain regions like the sgACC. Thus, we developed and validated a rsfMRI based personalization method for selection of left DLPFC targets. We used this personalization method to target 10 Hz rTMS and iTBS in healthy volunteers in a double-blind, sham-controlled crossover study design. We reported robust effects from single sessions of personalized 10 Hz rTMS/iTBS on the functional connectivity of the DMN. A single session of personalized 10 Hz rTMS disengages the sgACC and ventral striatum (vStr) from the DMN ~ 30 minutes after the stimulation. These effects dissipate over time and are not observed ~ 45 minutes after stimulation. On the other hand, a single session of personalized iTBS disengages the dACC, and rACC from the DMN ~ 30 minutes after stimulation. These effects become more intense ~ 45 minutes from stimulation, further disengaging rAI, and parts of MPFC in addition to dACC and rACC. For the first time, using the same study design, we depict that effect of respective rTMS protocols on the DMN connectivity diverge, spatially and temporally. Thus, our results support fundamental differences in the underlying DMN mechanisms of 10 Hz rTMS and iTBS. Notably, these effects were associated with personality trait of HA, hinting at plausible use of HA for predicting response to 10 Hz rTMS (low HA) or iTBS (high HA). We further translated this personalized approach to patients with depression by applying aiTBS at prospective sites and studied its effects against those resulting from standard F3 stimulation. This is the first instance of such a comparative study. We find a stronger decrease in sgACC-DMN functional connectivity in response to personalized iTBS than to F3 iTBS. Considering the reduced HRV in patients with depression, we further discover that HRV increases significantly during real iTBS only during personalized condition but not F3. The results further strengthen the premise that personalized approaches are more efficient in targeting deeper brain region functional connectivity than non-personalized approaches. Our results suggest larger benefits from use of a personalized approach for rTMS antidepressant treatment over the currently utilized F3 method of targeting rTMS.

7.1 Personalized rTMS target selection based on rsfMRI data

Our personalization method involves selection of left DLPFC sites based on their anti-correlation to DMN [163–165] and correlation to the network within which the target was located. The personalization method is robust, indicated by the higher negative correlation between individualized targets and sgACC than that between anatomical left DLPFC and sgACC. Thus, anatomical left DLPFC sites are less effective at addressing inter-individual variability in functional architecture of the brain. Furthermore, the personalization method is also reproducible. The test-retest assessment of the method in healthy subjects and patients with depression (data not shown) revealed that targets chosen based on rsfMRI data from the same subject, but different days are within 20 mm of each other (the diameter of focal region of figure-of-eight coil). Several other groups have also put forward rsfMRI based personalization methods. One such study explored target selection utilizing healthy human rsfMRI data. They used sgACC and left DLPFC ROIs to locate voxels in left DLPFC with maximal anti-correlation to the sgACC [269]. Their test-retest assessment revealed that targets chosen by their method laid within 25 mm of each other, much higher than our reported variability (10.9 mm). This is likely because their method relies on ROI definition which is sensitive to preprocessing steps like registration and normalization. In contrast, our method is based on large-scale networks, which can be robustly identified in non-normalized rsfMRI data. Another group used rsfMRI data to identify personalized left DLPFC targets by maximizing the difference between probability of a voxel belonging to DAN and DMN [247]. However, they did not report on test-retest assessment of their method. Hence, it is difficult to judge the clinical utility of their method. However, the use of rather long rsfMRI sessions (15 and 16.5 minutes respectively) is common across both these studies. Arguably, methods performing well with lower lengths of data would be favourable for an easier "bench-to-bedside" translation. In this aspect, our method is reportedly better at identifying personalized targets with as little as 5 minutes of rsfMRI data. Cole et al. [248] presented another alternative approach to personalization using just 8 minutes of rsfMRI data. They developed an algorithm to find targets for stimulation by determining the correlation between left DLPFC subunit and sgACC subunit while correcting for its size, and the spatial concentration of the subunit [248]. The test-retest assessment for their method is also absent. However, they report a significant response from applying aiTBS using their personalized approach in a sample of 21 patients with treatment resistant depression. In the absence of comparison of their novel method to a standard technique, the clinical superiority of their method remains to be explored in future studies.

The work presented here shows the feasibility of a rsfMRI based method of rTMS target personalization in both healthy subjects and patients with depression. The method is robust and reproducible for potential clinical application. By using this method for personalized stimulation of healthy subjects, we find effects in deeper brain structures using both 10 Hz rTMS and iTBS. The personalization method presented here provides an alternative to Deep rTMS for inducing more precise effects in deeper brain regions. The dispersed nature of the electric field induced by Deep rTMS [5] makes precise modulation of neural circuits difficult. Further, we employ the same method to delineate the differences in effects of a personalized versus standardized (F3) approach and find that the personalized approach is more robust at influencing deeper brain regions.

7.2 Default mode network changes after a single session of 10 Hz rTMS in healthy subjects

Here we applied for the first time a complete single dose of FDA approved 10 Hz rTMS in healthy individuals to understand the underlying mechanisms involving the DMN. We report statistically significant reduced functional connectivity of the sgACC to the DMN. To our knowledge, only one other study [228] attempted to delineate the mechanisms of 10 Hz rTMS in healthy subjects. However, they used a shorter protocol with less pulses, stimulated at anatomical left DLPFC, and analysed a network other than the DMN during a different time window. The differences between the previous study and ours include the target localization methods (anatomical vs. functional), networks analysed, observation time windows ([15-21 min] and [31-37 min](225) vs. [10-15 min], [27-32 min], and [45-50 min]), and rTMS protocol parameters (1200 at 80 % RMT vs. 3000 pulses session at 110% RMT) which could have collectively resulted in discrepant results between the two studies, thus making direct comparisons difficult. In addition to the sgACC, we also find that a single session of personalized 10 Hz rTMS decreases the functional connectivity of the vStr to the DMN. The vStr along with the nucleus accumbens (NAcc) contribute significantly to reward circuitry [270]. In patients with depression, studies have reported impaired bottom-up contributions and top-down control of NAcc [271] in addition to reduced functional connectivity of the NAcc to reward circuitry [272]. Furthermore, disruptions in the reward circuit also correlate to the severity of depression [272]. Considering the importance of sgACC and NAcc in depression pathophysiology, our results further strengthen the premise that personalized rTMS is successful at precisely influencing the functional connectivity of relevant deeper regions.

Additionally, we explored the temporal dynamics of effects from personalized 10 Hz rTMS to shed light on propagation of rTMS effects within and across large-scale networks. For this, we followed, along the four observation time windows, the connectivity between the stimulated DLPFC sites and the network that constitutes this region (IC-DLPFC), the sgACC-DMN connectivity, and the anti-correlation between the within network connectivity of stimulated site and sgACC. We find that the connectivity of left DLPFC to IC-DLPFC, showing a decrease, is the first to respond ~ 10 minutes after stimulation. This reduced connectivity, possibly a result of saturation of underlying neuronal population [22], starts to return to baseline during the later observation periods. The sgACC-DMN connectivity reduces only after ~ 30 minutes post stimulation and partially increases towards baseline levels ~ 45 minutes after stimulation. The differences in the observation windows during which effects are observed across left DLPFC and sgACC suggests a slow propagation of rTMS effects from stimulation sites to deeper brain regions. The changes observed in the anti-correlation, between the respective network connectivity of left DLPFC and sgACC, indicates shifting control of the IC-DLPFC over the DMN. Together, these findings implicate a plausible causal relationship between changes in functional connectivity of left DLPFC to IC-DLPFC and sgACC-DMN.

Aside from the sgACC role in depression pathophysiology, it also plays a crucial part in healthy individuals' response to sadness and anxiety. Previous works have shown increased blood flow to the sgACC in response to sad stimuli in healthy adults [256, 257]. We do not find any relationship between the decrease of sgACC-DMN functional connectivity and changes in negative affect on the PANAS. This is likely because associations between sgACC activity/connectivity and sad mood may be more conspicuous when actively engaging the region through tasks rather than when at rest. However, we aimed to explore large-scale network effects of rTMS rather than effects on singular regions, hence we did not employ any tasks to probe sgACC region's role in sad mood. Additional post-hoc tests, however, show a positive trend between the NAcc-DMN functional connectivity ~30 minutes after stimulation and a decrease in negative affect. This effect is not significant, likely because a single session of stimulation in healthy subjects would not result in drastic changes in negative affect (i.e. small effect size). Thus, the small range of changes observed in our sample might not be enough to delineate such relationships both from a biological and statistical viewpoint. However, the observed trend association between NAcc-DMN connectivity and changes in negative affect is interesting, considering the NAcc plays a crucial role in closed and open loops between cortical and subcortical brain regions that result in the expression of emotions and mood [273]. The NAcc region projects to the ventral tegmental area, vStr and receives inputs from amygdala and cortical regions. It has known roles in reward processing as well as avoidance of aversive stimuli [273]. Another study found a correlation between the ability to perceive emotions (using Mayer–Salovey–Caruso Emotional Intelligence Test) and the connectivity between the anterior DMN and the basal ganglia/limbic network (comprising the NAcc) [274]. They found that higher network connectivity was associated with lower scores on ability to perceive emotions. Thus, the lowered functional connectivity between the NAcc and the DMN observed ~30 minutes after stimulation may have modified the subject's perception of emotions. The explanation for why a modified emotion perception led to reduced negative affect perception rather than increased positive affect perception may be the accompanying reductions in sgACC-DMN connectivity.

Several studies have explored interactions between personality and brain functioning in healthy subjects to report a positive association of HA with activity in the sgACC [251], vmPFC-amygdala connectivity [275], as well as sgACC-amygdala connectivity [276]. We report that the changes observed in sgACC-DMN connectivity after a single session of 10 Hz rTMS in healthy adults were negatively correlated to the HA scores of the participants. This relationship suggests a possible role of HA scores in predicting 10 Hz rTMS effects. Thus, our results suggest robust effects of 10 Hz rTMS over DMN via sgACC and vStr, with a negative correlation between changes in sgACC-DMN connectivity and HA scores. This introduces the question if other clinically relevant rTMS protocols may have similar mechanisms. To answer this, we used the same personalization method for probing the DMN connectivity and its time lapse after a single session of iTBS. We chose iTBS, considering its recent FDA approval based on its non-inferiority to 10 Hz rTMS [88]. Delineating its DMN mechanisms can help understand the underlying effects of iTBS as well as its differences from 10 Hz rTMS, with potential application for clinical decision making.

7.3   Default mode network changes after a single session of iTBS in healthy subjects

Application of a single iTBS session (1800 pulses) in healthy subjects reveals that the effects of iTBS are first observed as a decreased connectivity between nodes of the SN and the DMN. Furthermore, these effects are dynamic as revealed by following the changes through multiple time windows. Approximately 27-32 minutes after stimulation, we report the first changes in DMN connectivity to the rACC and the dACC. With time, the magnitudes of these changes strengthen, and the effects propagate to other nodes of the SN (rAI), as observed 45-50 minutes post stimulation. These results strongly suggest that in healthy adults iTBS does not engage with the DMN directly via its own nodes but exerts its influence on the DMN via the nodes of the SN. An analysis of the temporal dynamics of changes in the rACC to DMN connectivity and stimulated DLPFC to IC-DLPFC connectivity suggests a possible feedback of rAI induced perturbations into the DLPFC causing its connectivity to alter during later periods of observation. However, more future work is needed to further explore this hypothesis. Previous studies with iTBS in healthy individuals have reported different iTBS effects. The effects from a 3-minute iTBS protocol on an anatomically located left DLPFC, revealed that immediately after iTBS functional connectivity of the left dorsolateral superior frontal gyrus (SFG) to the left dorsal inferior frontal gyrus (IFG), and left rostral IFG to the right middle frontal gyrus (MFG) increased while that between the left orbital gyrus (OG) to the left lateral OG decreased [277]. Approximately 15 minutes after the stimulation, left IFG-right amygdala and bilateral IFG-OG functional connectivity decreased while that between the left MFG-left OG increased [277]. Another study applying the identical protocol at an anatomical definition of the left DLPFC found increased connectivity between the left DLPFC and the bilateral caudate when iTBS was delivered at 90% RMT [278]. However, iTBS at 120% RMT resulted in increased right amygdala – right caudate connectivity. The lack of increased left DLPFC connectivity when iTBS was delivered at 120% RMT suggests a possible role of homeostatic processes to reduce connectivity when stimulated above a certain threshold [278]. These studies indicate varied connectivity changes from iTBS the over left DLPFC. However, the use of 3-minute iTBS at an anatomically localized left DLPFC while using seed-based analysis, compared to 10-minute iTBS at functionally defined left DLPFC and ICA-based analysis used in our study, makes relevant comparison difficult. One study however, did apply 3-minute iTBS over functionally relevant left DLPFC sites chosen based on their connectivity to the rAI [229]. Using GCA, they found a decreased functional connectivity of the DLPFC and insula after stimulation. This finding along with our report suggest iTBS at functionally well-defined left DLPFC nodes can perturb the functional connectivity of rAI.

Another important aspect of the report by Iwabuchi et al. is that they observed the perturbation of left DLPFC-rAI connectivity as early as 20 minutes after iTBS [229]. Our neuroimaging results suggest that reduction in rAI-DMN connectivity is only observed ~45 minutes after stimulation. This may be a consequence of difference in iTBS session length. Alternatively, choice of left DLPFC sites based on their anti-correlation to sgACC may result in longer time needed for effects to be observed in rAI. We further explored the temporal dynamics of iTBS induced changes in the functional connectivity of left DLPFC to IC-DLPFC, rACC-DMN, and the anticorrelation between within network connectivity of left DLPFC – [IC-DLPFC] and sgACC-DMN. While rACC-DMN connectivity follows changes reported in the neuroimaging results, the left DLPFC-[IC-DLPFC] connectivity does not change until ~45 minutes after iTBS. This contrasts with 10 Hz rTMS, where a change in left DLPFC-[IC-DLPFC] connectivity was observed as soon as ~10 minutes after stimulation. This divergence indicates that effects of iTBS may not be mediated by changes in connectivity of the left DLPFC within its own network, although connectivity of the left DLPFC with other individual regions may change in response to iTBS, as reported elsewhere [229, 278, 279]. Thus, the influences observed in the dACC, rACC, and rAI may be mediated by other brain networks/regions. The changes observed in the anti-correlation of the left DLPFC-[IC-DLPFC] connectivity to the sgACC-DMN connectivity indicate iTBS induced changes in IC-DLPFC control of the DMN, as observed after 10 Hz rTMS. Thus, the reduced connectivity of the dACC, rACC, and rAI may be mediated by DMN nodes without influencing the connectivity of such nodes within the DMN. This may be because stimulation sites were chosen based on their anticorrelation to the DMN. Thus, the lack of effect in left DLPFC-[IC-DLPFC] connectivity up until ~45 minutes after stimulation is likely a result of concurrent reduction in rAI-DMN connectivity. Considering influences of iTBS modulation on left DLPFC-rAI connectivity [229], we hypothesize that the decreased rAI-DMN connectivity causes a corresponding increase in the rAI influence on the DLPFC, which further results in the increased left DLPFC-[IC-DLPFC] connectivity observed ~45 minutes after iTBS. Finally, the correlation between the left DLPFC-[IC-DLPFC] connectivity and the sgACC-DMN connectivity is maximal until ~45 minutes after stimulation. Given the strongest effects are also observed during the last time window, the iTBS effects are likely to last beyond our observation window of 50 minutes.

One advantage here is the use of identical study designs for exploring the underlying mechanism of the DMN in response to either 10 Hz rTMS or iTBS. We have demonstrated that both the spatial effects and temporal dynamics of such effects are different for 10 Hz rTMS and iTBS. In the case of iTBS, we did not see any correlation between affect measured by the PANAS and the changes observed in functional connectivity, unlike after 10 Hz rTMS. This is likely because of different regions affected by each of the rTMS protocol, with iTBS effects observed in nodes other than the sgACC and NAcc. The lack of involvement of the NAcc and sgACC (regions implicated in emotion regulation and anhedonia) in response to iTBS, might explain why we do not observe such effects in the iTBS study. Furthermore, we report that changes in rACC-DMN connectivity are positively correlated with HA scores. This relationship contrasts with that observed from 10 Hz rTMS where the rTMS induced sgACC-DMN connectivity was negatively correlated with HA. Arguably the opposite results observed in the experiments can be attributed to differences in regions whose connectivity to the DMN is affected in response to either 10 Hz or iTBS. Thus, healthy subjects with higher HA, presumably have higher rumination, and anxiety and vigilance towards external stimuli. Such individuals are more likely to respond to iTBS via disengagement of SN nodes from the DMN. On the other hand, subjects with lower HA scores show lower disposition for rumination, anxiety, and vigilance towards external stimuli and are more likely to respond to 10 Hz rTMS. Considering the opposite relationship between HA and DMN effects from 10 Hz rTMS and iTBS, we speculated that HA scores may facilitate personalizing rTMS protocols based on personality types. To our knowledge, this is the first study using identical study design to follow DMN effects for up to 50 minutes after stimulation with either of the two clinically relevant rTMS protocols. However, these results are based on healthy individuals and their generalizability to patient cohort is limited. Thus, we further investigated the DMN mechanisms of personalized rTMS in patients with depression. As patients with depression have higher HA than healthy controls [184], and we saw a positive relationship between HA scores and iTBS induced changes in healthy subjects, we expected that iTBS may be more beneficial for antidepressant treatment. Additionally, the shorter duration of the iTBS protocol reduces the duration of treatment sessions without compromising treatment efficiency [88], presumably reducing the extent of scalp discomfort (shorter duration implies shorter periods of discomfort). We hence chose 10-minute iTBS protocol to be applied in an accelerated manner to study the DMN mechanisms in patients with depression. The accelerated protocol reduces the treatment duration from four weeks to one, expectedly aiding in patient compliance to complete the full duration of treatment.

7.4 Default mode network and cardiac changes after multiple sessions of iTBS in moderately to severely depressed patients

We applied aiTBS in patients with depression using either our rsfMRI based personalization method or the EEG 10-20 system based F3 method of targeting the left DLPFC. We aimed to investigate the effects from personalized aiTBS as well as qualitatively compare its effects to non-personalized aiTBS. To our knowledge this is the first study to compare comprehensive effects (therapeutic, large-scale network, and autonomic regulation) from both personalized and non-personalized (F3) neuromodulation using the same design. Through preliminary analysis, we show that aiTBS reduces the connectivity of the sgACC and precuneus/PCC regions to the DMN. For the first time we report that this effect is stronger in the sgACC when a personalized approach is used as compared to an F3 approach. In line with stronger influences on the DMN in the personalized group, we also provide evidence using HRV that such a personalized approach is more effective at regulating HRV than F3. In accordance to the role of DMN nodes in both self-directed thought processing and autonomic regulation [221], we present evidence indicating that sgACC-DMN connectivity is associated with heart rate regulation. We also show that BDI-II scores are negatively correlated to DMN connectivity of SN nodes, and to HRV. Unlike the differential effects of personalized and non-personalized aiTBS on the DMN and HRV, we do not find any differences in the symptom improvement (BDI-II, HAMD, and MADRS). This indicates that while personalized aiTBS modulates the DMN and HRV more effectively than F3 aiTBS, these effects do not translate to greater therapeutic benefits. It is possible that the effects of precise modulation of the DMN on behaviour may not be captured by full symptom scales and analysis of more specific sub-symptoms may delineate the therapeutic benefits of personalized aiTBS better. Furthermore, the therapeutic response is a result of influences from aiTBS as well as the underlying heterogeneity of our sample. Future analysis controlling influences from confounder variables will further clarify the differential therapeutic effects of personalized and F3 aiTBS. Similar to earlier reports showing mood improvements after rTMS treatment in patients with depression [280–282], we also find a significant decrease in the negative affect measured using the PANAS daily. However, the negative affect reduces after sham stimulation too. Using the visual analogue scores (VAS) we found that patients' expected outcomes for their mental states from real and sham were the same even though the physical sensation from real iTBS were much stronger than the sham. Thus, considering that patients internally expected similar benefits indicates that the reported changes in mood is likely a placebo effect rather than an effect arising directly from iTBS itself.

Our evaluation of DMN effects arising out from aiTBS provides an opportunity to understand the differences between personalized and F3 iTBS in patients with depression. We find that personalized iTBS is more effective in reducing the sgACC-DMN connectivity during week 3 compared to week 1. In line with our initial hypothesis, we present evidence that precise stimulation of targets in left DLPFC based on their connectivity to sgACC, allows enhanced influence on the sgACC-DMN connectivity compared to the current standard practices, such as the F3 method. However, in both iTBS groups, a reduction in FC of the precuneus/PCC regions is seen in week 3 compared to week 1. This indicates that these effects are likely a result of dispersed effects of iTBS rather than precise modulatory effects arising from personalized aiTBS. Our observations of reduced sgACC-DMN connectivity in our cohort of patients with depression follows previous works which show that treatment of depression is accompanied by correction of the sgACC activity and connectivity aberrations [146, 268, 283]. Other regions showing a reduced functional connectivity to the DMN in our study, namely the precuneus/PCC regions, are also implicated in the pathophysiology of depression [146, 284]. Another recent work applying iTBS to the dACC in patients with depression found an increased connectivity between the stimulated target and the precuneus regions in relation to symptomatic improvement [285]. There are crucial differences between this study and ours in terms of target node (dACC versus left DLPFC) and manner of iTBS delivery (accelerated versus non-accelerated). While recognizing this, we hypothesize that an increase in connectivity of the precuneus to the dACC (SN node) in response to iTBS may occur concurrently with a decrease in connectivity of the precuneus to the DMN as seen in our study. It is important to note that the regions showing altered connectivity to the DMN in our healthy cohort (SN nodes) are not the same as those showing alterations in patients with depression (DMN nodes). This suggests that mechanisms of multiple sessions of iTBS are different under depression pathophysiology as opposed to effects from a single session in healthy brains. However, the connectivity of the DMN to these SN nodes plays a role in mediating depression severity evidenced by the negative relationship between the depression severity (BDI-II) and the functional connectivity of the DMN to the dACC, rACC, and the left AI. A similar negative relationship has been noted earlier between rAI-SN connectivity and BDI-II scores [142]. Our results hence further support the role of left AI connectivity in mediating depression severity. The dACC plays a crucial role in conflict resolution and cognitive control while the AI is important for social and affective information [141]. The reduced rAI-SN connectivity [142] may result in poor switching between the DMN and CEN roles, resulting in ruminative behaviour (larger symptom severity), while greater dACC-DMN and left AI-DMN connectivity may translate into larger cognitive control over processes mediated by the DMN, such as rumination (lower symptom severity). Finally, the lack of significant relationship between the connectivity of the DMN with SN nodes and symptom changes suggests that these regions may only mediate depression severity but not symptomatic improvement.

We provide further evidence of precise modulatory effects from personalized aiTBS through analysis of HRV influences. Our results suggest personalized aiTBS is more successful in increasing HRV and trend of reduced HR than F3 aiTBS. This is likely a result of poor target engagement by the F3 method, which results in minute effects (higher slope in the first 45 seconds of stimulation start) that do not translate to changes in HRV observable over the larger 10 minutes duration of iTBS. Our findings support those from a previous report showing 3-minute iTBS over anatomically defined left DLPFC in patients with depression resulted in an increased HRV and reduced HR [220]. We also reported that in the personalized aiTBS group, the slope of the RR interval time series shows a negative association to sgACC-DMN connectivity. This relationship is not observed for the F3 aiTBS group likely because of an absence of more robust alterations in sgACC-DMN connectivity. The relationship between the slope of the RR interval time series and the sgACC-DMN connectivity implies that a higher sgACC-DMN connectivity is associated with reduced parasympathetic influences on the heart (tendency for heart rate acceleration). The results are similar to recent findings in marmosets showing a similar relationship between increased sgACC activity and reduced parasympathetic influences on HR [222]. Our findings further support previously reported links between the DMN role in self-directed thought process and autonomic functioning of DMN nodes [192, 221]. Our report of an inverse relationship between HRV and depression severity are in line with previous results [196, 205, 286, 287]. Such a relationship indicates that increased depression severity is associated with further reduced HRV, depicting the damaging effects depression pathophysiology has on autonomic functioning.

7.5 Limitations

We employed the VAS to explore the perception of scalp discomfort and expected effects on mental states in both healthy subjects and patients with depression. We found that blinding integrity was maintained during the application of personalized 10 Hz rTMS in healthy subjects, as well as aiTBS in patients with depression. However, when applying a single session of iTBS in healthy subjects, we found that the expected mental state effects were different for real and sham conditions. Further exploration of the DMN changes associated with only the real iTBS without correcting for sham effects show that the same nodes (dACC, rACC, and right AI) show connectivity changes to the DMN. This indicates that effects observed are not driven by sham stimulation but are specific to those caused by real iTBS. The temporal exploration of sham effects on the functional connectivity of stimulated DLPFC-[IC-DLPFC] and rACC-DMN connectivity revealed no changes analogous to that observed from real iTBS. This suggests that the results are indeed affected by real and not sham iTBS. Some subjects who participated in the iTBS study were also prior participants of the 10 Hz rTMS study. It is possible that prior experience with an rTMS stimulation might have resulted in distinct expectations in further sessions. The small sample size of the preliminary analysis of patients with depression is a notable limitation of the study. Of particular importance is that we do not find significant differences in therapeutic effects from personalized and non-personalized iTBS. This is interesting, considering we noted more robust effects of personalized aiTBS on both the DMN connectivity and HRV. As the larger effects of personalized rTMS are seen in sgACC connectivity, we hypothesize that symptoms mediated by sgACC functioning may be the ones to display differences between personalized and non-personalized rTMS. Hence, future analyses must also attempt to study the sub-types of depression symptomatology to detect effects within them. Also, the underlying heterogeneity in the sample (medication loads, age, gender, etc.) could have influenced the therapeutic effects from aiTBS. Thus, analysis of larger sample sizes while controlling for confounders, will further clarify if absence of larger therapeutic benefit was an effect of a lack of power in our study. We also do not observe any relationship between symptomatic changes and HA scores, as we hypothesized. This too may be an outcome of small sample size of our preliminary analysis. Some studies however, indicate that rTMS effects in patients with depression are associated with personality [187, 188]. Thus, future analysis with larger sample size will clarify the role of personality in predicting aiTBS antidepressant response. Also, considering the potential implication of pharmacological antidepressant effects on HA, the relationship between HA and symptom improvement may have been confounded by such additional variability. It is important to mention that we see differential DMN and HRV effects from personalized and non-personalized aiTBS, despite the small sample size. This nevertheless suggests likely differences between personalized and non-personalized aiTBS that future analysis will help establish.

7.6 Outlook

Through our work, we present a novel approach to personalize rTMS for the treatment of depression. We validated the usability of this new approach in both healthy individuals and patients with depression. Our work with healthy volunteers is instrumental in highlighting the spatial and temporal differences in DMN effects arising from both iTBS and 10Hz rTMS protocols. Owing to its much shorter duration and comparable therapeutic efficiency to 10 Hz rTMS, we consequently chose the iTBS protocol for exploring the effects on the DMN in patients with depression. We simultaneously compared these effects against those resulting from a non-personalized method of targeting. As hypothesized, when selecting rTMS targets based on their connectivity to the sgACC, we see a more pronounced effect on sgACC-DMN connectivity within the cohort receiving the personalized protocol. The precise nature of such modulation is also observed through ECG data, showing that personalized stimulation is more successful in increasing HRV parameters than non-personalized (F3) stimulation. As the HRV indices are lower in patients with depression, these results suggest personalized stimulation may be more beneficial than the F3 based approach in alleviating HRV based deficits. Attention to parameters such as HRV deficits could help to further support the advancement of personalized approaches in clinical practice.

Given our interest in understanding and developing a personalized rTMS approach specific to the treatment of depression, our evaluation has been confined to the DMN. Analysis of large-scale networks aside from the DMN will shed further light on comprehensive responses of brain networks to multiple sessions of aiTBS. Use of data-driven approaches such as multivariate analysis, deep learning, etc. will allow delineating the linear as well as non-linear interactions between large-scale networks and how these interactions change in response to neuromodulation. Such works will also help aid in the development of more accurate biological models of depressive disorders and promote a better understanding of the therapeutic effects of treatment interventions.

Our work has only explored a singular aspect of a personalized rTMS approach, i.e. the target of stimulation. However, rTMS protocols are highly variable and the stimulator parameters alone represent a field of study wherein the determination of the ideal set of parameters for each person's stimulation are ongoing. Furthermore, the physiological state of an individual's underlying cortex during the stimulation can influence the stimulation outcome. The physiological state can be influenced by biological, psychological and environmental factors. Hence, it is an avenue that should be explored by controlled studies examining the concurrent physiological state of cortex being stimulated and how it influences rTMS outcome. Furthermore, additional sources of variability that affect the therapeutic rTMS response are diverse, ranging across phenotypic, genotypic, and connectomic domains. Thus, further steps towards a personalized medicine approach requires not just revealing the influences of these factors on treatment outcome, but also developing an integrated framework wherein all sources of variability can be evaluated. Together, these results can help to guide clinicians to the most beneficial treatment modality for a given individual.